50 ways
to start a
garden

RHS 50 Ways to Start a Garden
Author: Simon Akeroyd
First published in Great Britain in 2022 by Mitchell Beazley,
an imprint of
Octopus Publishing Group Ltd, Carmelite House,
50 Victoria Embankment,
London EC4Y 0DZ
www.octopusbooks.co.uk

An Hachette UK Company
www.hachette.co.uk

Published in association with the Royal Horticultural Society
Copyright © 2022 Quarto Publishing plc

ISBN: 978-1-78472-844-1

A CIP record of this book is available from the British Library
Set in Garamond and Futura
Printed and bound in China

Mitchell Beazley Publisher: Alison Starling
RHS Publisher: Rae Spencer-Jones
RHS Consultant Editor: Simon Maughan
RHS Head of Editorial: Tom Howard

Conceived, designed and produced by
The Bright Press
Part of the Quarto Group
The Old Brewery, 6 Blundell Street, London, N7 9BH, England
www.Quarto.com

Publisher: James Evans
Managing Editor: Jacqui Sayers
Editor: Sara Harper
Project Editor: Katie Crous
Design & Picture Research: Wayne Blades
Senior Designer: Emily Nazer
Illustrations: Sarah Skeate

The Royal Horticultural Society is the UK's leading gardening
charity dedicated to advancing horticulture and promoting
good gardening. Its charitable work includes providing expert
advice and information, training the next generation of
gardeners, creating hands-on opportunities for children to
grow plants and conducting research into plants, pests and
environmental issues affecting gardeners.

For more information visit www.rhs.org.uk
or call 0845 130 4646.

RHS

50 ways to start a garden

IDEAS AND ADVICE FOR GROWING INDOORS AND OUT

SIMON AKEROYD

MITCHELL BEAZLEY

CONTENTS

INTRODUCTION

50 Ways to Start a Garden is here to help and inspire you as you take those preliminary steps into the exciting world of horticulture. Whether you have inherited an overgrown jungle or are staring at a blank canvas; whether you have a small or moderate-sized garden outside or just space inside for a few house plants, starting your own garden will make your space look better and feel more relaxing and inviting. Furthermore, surrounding yourself with plants, all pumping out oxygen and soaking up carbon dioxide, will provide you with cleaner air and, therefore, improve your physical health and mental wellbeing, too.

Below: Foliage plants are a wonderful addition to an indoor garden, providing an exciting range of colour, texture and height.

Getting started, you will learn how to assess your growing space so that you can choose appropriate plants for it. There is also a section on which tools you will need, and you'll be introduced to a selection of different design or planting styles, so you can pick the ones with maximum appeal and that fit with your lifestyle. Some suggestions are quick fixes requiring very little maintenance, whereas others require more time and investment.

There are 50 ideas to incorporate in your gardening – including information and projects for both indoor and outdoor gardens – from adding height to creating a no-dig garden; from making a dry garden to installing a rockery. There are also some handy step-by-step projects that are easy to follow and require no or little DIY and/or gardening skills, such as creating a wildlife pond and weaving a willow screen. In addition, there are a number of plant profiles, featuring some of the best species of certain groups of plants that you might want to start growing, such as herbs, cacti, alpines, vegetables and fruit.

So, what are you waiting for? Pull on your wellies, don your gardening gloves and get stuck in.

Assess your growing space

Creating a garden is exciting, as well as providing an opportunity to express your personality. It might be tempting to get stuck in straight away, but before getting started, it is important to assess the site. This will help you make the best choices about what to grow and where.

If you have space to grow plants outside, there are a few questions to ask yourself before starting, no matter how big or small the plot. Is the garden simply somewhere to sit outside and enjoy the surroundings, or do you want to get your hands dirty and enjoy the thrill of growing your own plants? How much time to you have to look after your garden? Should it be easy to care for, or do you have enough time on your hands for more complex elements? If you live a busy life, opt for something minimal, such as a gravel garden, a relaxed seating area and a few seasonal drought-resistant plants in containers (see Design a no-fuss garden, page 58).

There are some practical considerations when starting to grow plants. If growing outside, consider issues such as where the bins will go, whether there is space for a compost heap and where the tools will be stored. If growing indoors, you will still need space to store a few basic tools (see page 12). You should also ensure the plants are not going to be knocked or damaged, or are not going to have to be moved every time you need to open a cupboard door or window. Consider the speed at which plants grow, because some may need more care or maintenance, such as frequent repotting or trimming. If you want low-maintenance house plants, choose ones with slow-growth habits.

What's your style?

This is an opportunity to be creative. Would you like your outdoor space to be formal or informal? If the former, then you may want to incorporate straight design lines, clipped hedges and geometrical shapes, and include features such as topiary, sculptures, mowed lawns and symmetrical planting schemes. Informal features have a more relaxed feel to the overall design, which could include wildlife areas, mixed planting, rambling roses over arches and lots of plants jostling with one another.

There may be other themes that could be emulated such as Japanese (see page 156), dry gardens/Mediterranean (page 34), subtropical (page 38) and potagers.

Opposite: Even in tiny courtyard gardens there are plenty of opportunities to surround yourself with plants, in containers or trained up walls and fences.

Once you have considered the above, draw up a plan to identify the location of paths, storage areas, compost heap, pots and planters, and the overall structure.

Light and shade

It's important to assess the growing space for the amount of sunlight it receives during the day, so that you can put your plants in the best location to suit their needs. In the northern hemisphere, the sun rises in the east and sets in the west, so usually a south or southwest-facing garden, windowsill or balcony will receive the most sun. Anything facing north may not receive any direct sunlight.

The great thing about plants is that there are plenty suited to all sorts of different conditions ranging from deepest shade to brightest sunlight, so choose plants accordingly and they should thrive.

If growing plants on a windowsill, be aware that the glass magnifies and intensifies direct sunlight, so be careful that foliage and flowers do not burn, and the compost does not dry out.

The answer lies in the soil

Most plants have a specific type of soil they require to survive. If growing indoors, plants are almost always grown in a type of potting compost, so it is easy to provide the correct conditions. Simply check the label or look up the plants' requirements in a book or online, and choose an appropriate compost. Also check watering requirements, as many house plants do not like overwatering.

When growing outside, there are a few more variables regarding a plant's growing medium. Some prefer light, sandy soils with low fertility, others prefer something heavier with more nutrients. Always read the plant label when buying from the garden centre, to see what the soil requirements are.

Clay soils are usually fertile but can be hard to dig and have poor drainage, becoming water-logged in winter. Sandy soils are light, low in fertility and free-draining, so plants may need more watering and feeding than in clay. Plants that prefer drier conditions or are drought-tolerant are best suited to this soil. Adding organic material such as garden compost or manure can improve the conditions for both clay and sand.

Right: A fine late-summer display can be made of fiery heleniums with veronicastrums, Chinese astilbes and grasses, as they all tolerate the same range of soil and light requirements.

Soil test

It may be worth doing a pH test to find out your soil type. Some plants require an acid soil, often referred to as ericaceous. This includes rhododendron, camellias and blueberries. If the soil in your garden is not on the acid side of neutral (pH 7), grow acid-loving plants in containers, in ericaceous compost. Soil-testing kits are available online or in garden centres.

Get the right equipment

There are a few gardening items worth having in your armoury when caring for plants. The equipment required will depend on the size of your growing area and what type of plants you intend to grow. Indoor gardening items can usually be stored in a cupboard, but for larger gardens you may need a secure shed. These are not complete lists, but will certainly be enough to get you started.

Indoor gardening

Thankfully, most tools required for indoor gardening are small, making storage easy. They are also relatively inexpensive to purchase.

- **Watering can**: to give your plants the right level of moisture.

- **Saucers or trays**: to catch any drips.

- **Attractive pots**: to grow the plants in.

- **Crocks**: to place in the bottom of pots to improve drainage.

- **Secateurs**: to trim or snip shoots.

- **Compost**: choose one that is suitable for your selected house plants.

- **Labels and pencil**: note and keep track of which plants you have.

Small courtyard garden/small beds

Small gardens only need a handful of essential tools, and mechanized equipment is rarely needed.

- **Border spade**: smaller than a usual spade, great for small spaces.

- **Border fork**: for loosening soil and/or mixing in compost.

- **Mini compost bin or rotating drum:** for easy disposal of your kitchen waste.

- **Hand fork and trowel**: for weeding small areas.

- **Water butt and/or hose:** to make watering your garden as easy as possible.

- **Trug**: for collecting weeds or harvesting crops.

- **Secateurs**: for trimming shrubs.

Crocks

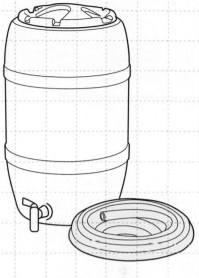

Water butt

Secateurs

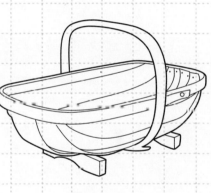

Trug

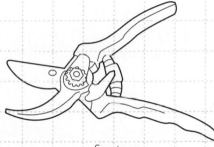

Pots and planters

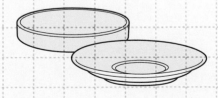

Saucers and trays

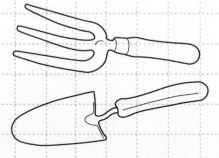

Hand fork and trowel

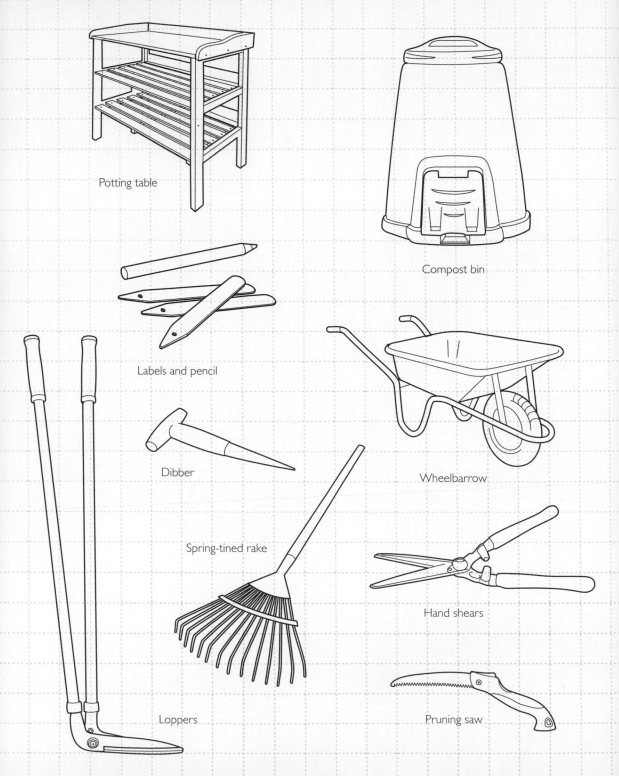

Potting table

Compost bin

Labels and pencil

Wheelbarrow

Dibber

Spring-tined rake

Hand shears

Loppers

Pruning saw

For propagating your own plants

Growing your own plants from seeds or cuttings will save you money, and only requires a few basic tools.

- **Table or potting bench**: for sowing.

- **Dibber**: to make holes for seeds in compost.

- **Watering can.**

- **Seed trays, pots and sowing compost.**

- **Labels and a pencil**: to remember what you have sown where.

- **Shelves or windowsills**: for putting trays of seeds on.

Lawns

Lawns require their own specialist equipment, with a mower probably being the most expensive but useful item to purchase.

- **Mower to cut the grass**: can be electric, battery-operated or a push mower.

- **Edging shears**: for trimming the grass edges.

- **Compost bin**: (ideally) for putting grass clippings into.

- **Spring-tined rake**: for scarifying the lawn.

- **Fork**: for aerating the lawn occasionally.

Medium-sized garden

Equipment gets more expensive as gardens get larger, but it is worth investing in quality products as they will be effective and last a long time.

- **Spade**: for planting, digging and moving soil.

- **Fork**: for cultivating the ground.

- **Wheelbarrow**: for moving material around the garden.

- **Compost area**: for recycling kitchen and garden waste.

- **Water butt and/or hose**: for watering plants.

- **Rake**: for levelling the soil.

- **Hoe**: for removing weeds.

- **Hand shears or a hedge trimmer**: if you have any hedges to cut.

- **Secateurs, loppers and pruning saw**: to cut shrubs and small trees.

(plus equipment from small courtyard garden/ small beds, see page 12)

Keep it clean

Always clean hand tools after they have been used by brushing off the dirt and wiping steel or metal blades with an oily rag.

3

Learn to water your garden

Water is the key to life everywhere, and it is, therefore, an essential element for a thriving garden. Unless you have selected specifically drought-tolerant plants, chances are that your plants will need watering occasionally. Watering correctly is essential for a plant to survive, and there are a few techniques to doing this successfully.

When to water

The most efficient time to water plants is in the morning, before it gets too warm. This is because plants can absorb the moisture in the soil as the day starts to warm up. It also means that the plant's foliage won't stay damp for too long, which can cause problems such as mildew and attract slugs. If early morning isn't possible, then the evening is the next best time, as the water will soak into the soil, ready for plants to use the following morning. Avoid watering during the hottest part of the day as the water will simply evaporate.

Where and how much to water

Water the soil at the roots of a plant, avoiding splashing the leaves, which can cause sun scorch and mildew problems. The amount of water needed depends on the size and type of plant, but as a rough guide, if a plant is in a container, then give it about a tenth of liquid to the size of the pot.

Rainwater harvesting
Attach water butts to downpipes from house gutters, and fix gutters to sheds and greenhouses to help collect rainwater. Rainwater is natural, free water for a plant.

For example, a 10-litre (20-pint) pot could be given 1 litre (2 pints) of water.

How often to water

Little and often is the key for pot-grown plants. For plants in the open garden, the opposite is true. A good soak in the morning/evening is better every now and again, as it encourages roots to grow deeper, which as a result will be more resilient to drought. The best way to tell if most plants need watering is to stick a finger in the soil. If it feels dry, then give it a drink. In the summer, containers, hanging baskets and young plants will need watering almost every day. In the winter, only water if the compost/soil feels very dry.

Overwatering in winter can cause plants to rot. If a plant is wilting or looking limp, then it almost certainly needs watering, although ideally it shouldn't reach this condition.

Water-retaining tricks

For pot plants, placing pots in trays or saucers is helpful, as it catches excess water run-off and prevents wastage. When the tray/saucer is empty, your plant needs a drink.

Adding organic matter to the planting hole helps to preserve moisture and reduce a plant's chances of suffering in hot and dry weather conditions. Mulching the surface area around the root ball will help reduce water loss through evaporation.

A bund can be made around a plant by mounding up the soil to about 5cm (2in) high in a circular shape about 20cm (8in) from the stem. This will prevent the water running off the surface of the soil and away from the root area, where it is needed most.

Above and opposite: Use a watering can where possible instead of a hose pipe, as this helps conserve water by focusing on where it is needed, with less splashing.

Plant bulbs to make the most of spring

Nothing heralds the end of winter and the start of a new season like spring bulbs pushing up from the ground, or trees covered in beautiful blossom. No matter what size of garden, there is always room for some seasonal interest at this time of year. Plant in autumn to reap the rewards in spring.

Bulbs in containers

If you don't have much growing space outside, then you can plant bulbs in containers. This also makes it easy to ring the changes, as when one sequence of bulbs finishes flowering, another one can be moved in to take its place. Plant bulbs in pots using peat-free general-purpose compost at two to three times their depth. Make sure the containers have drainage holes to avoid any rot. After the plants have finished flowering, plant them out in the garden to enjoy their display the following year.

Bulbs in the garden

Spring-flowering bulbs should ideally be planted in autumn if they are to flower the following spring. Take care to plant bulbs the right way up, with the flatter root plate at the bottom and the pointed tip facing upwards. Dig individual holes for the bulbs with a trowel, or if there are a few, it might be worth investing in a long-handled bulb planter, which can save on back-breaking work. If you want a dense cluster of flowers, simply dig one large hole or pit at the correct depth and fill it with bulbs.

Bulbs at the base of trees

Bulbs at the base of trees brighten up a garden in early spring, before the season warms up and other areas, such as flower borders, erupt into colour. The bright flowers from the spring bulbs and the attractive upright tree trunks complement and accentuate each other's qualities.

Choose trees with attractive trunks to act as a foil against the colourful flowers, such as white Himalayan birch (*Betula utilis* subsp. *jacquemontii*) or cider gum (*Eucalyptus gunnii*). Alternatively, consider Tibetan cherry (*Prunus serrula*), with its deep mahogany colour, or a snake-bark maple, such as *Acer davidii*, with its green and white serpentine stripes, which will contrast with the bright flowers of crocus and cyclamen.

Clockwise from top left: Enjoy a recurring floral display with spring bulbs grown in containers.

The yellow of daffodils and the blue of grape hyacinths are two classic, contrasting spring colours.

Cyclamen make a bright splash of colour in the dappled shade of trees.

Bulbs come in a range of sizes, with large ones needing to be planted deeper than smaller ones.

SPRING-FLOWERING PLANT SELECTOR

Spring-flowering bulbs are a quick and easy way to create a vibrant splash of colour in the garden before the season warms up and gets fully underway. There are hundreds of different ones to choose from in a range of colours and sizes. Some can be used for fragrance; others in borders. Bulbs are also useful for displays in containers, if space is at a premium.

DAFFODILS
Narcissus

These popular yellow flowers with their nodding, trumpet-like heads create a bold statement in lawns, borders and around the base of trees. There are a few different types, including the larger hybrid ones or the smaller fragrant species types such as 'Tête-à-tête', which only grow to about 15cm (6in) high.

TULIPS
Tulipa

If you want a riot of colour in your border, then these are the bulbs to go for, although there are more subdued species types, too. Different forms of tulips will flower at different times, so with a bit of planning it is possible to enjoy a succession of colours from early to late spring.

HYACINTH
Hyacinthus

If you want a sweet fragrance to pervade your entire garden during spring, then grow hyacinths. They come in mainly blue, white and pink, and have large attractive flowerheads. Cut flower stems back when they have faded and they should produce more flowers the following year.

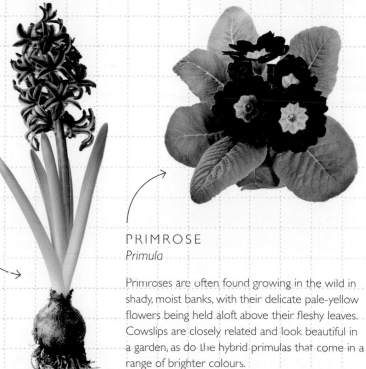

PRIMROSE
Primula

Primroses are often found growing in the wild in shady, moist banks, with their delicate pale-yellow flowers being held aloft above their fleshy leaves. Cowslips are closely related and look beautiful in a garden, as do the hybrid primulas that come in a range of brighter colours.

SNOWDROPS
Galanthus

Known as the harbingers of spring, these delicate flowers with their nodding white heads can make appearances from mid-winter onwards. They prefer moist yet well-drained soil, and are best naturalized in lawns, in partial shade.

CROCUS
Crocus

These low-growing bulbs can provide a range of colours in the garden including purple, cream, orange and yellow. They are often best when naturalized in the lawn, but can make a great display in containers, too. There are also ones that flower in autumn or winter, so choose carefully if you are wanting a spring display.

5

Plant a wildlife-friendly garden

One of the many joys of having a garden is sharing it with all the abundant wildlife around us. Just a few containers or a single window box full of flowers will attract wildlife with their aromas and colours. Even a dish of water will attract birds, bees and butterflies, as they rest and quench their thirst when going about their business.

To attract a wide range of wildlife, it is best to have a selection of different habitats with as many different types of plants as possible within your garden. Consider plants that will flower at varying heights, to accommodate all the different-sized creatures that may use the garden. Include trees, shrubs, herbaceous and annual flowers, and low ground cover. Climbing plants such as ivy can provide a late source of nectar, while the fluffy flowers of old man's beard (clematis) can be used by birds for nesting material.

All-year-round interest

Not only is it more interesting for humans to have all-year-round interest in the garden, but wildlife will benefit, too. Birds, bats and insects will reward you with regular visits, or even by making your garden their home, if you create a garden with plants flowering or fruiting at different times of the year. After all, if you wish to retain wildlife, they will need to be able to find food, drink and nesting material all year round.

Right: Plant a wide range of nectar-rich flowers, such as this buddleja, to encourage butterflies into the garden.

Opposite, from top: Berrying plants such as rowan will encourage feeding birds to visit.

Place a shallow dish with water in the garden for birds to bathe in and drink from.

Not only do water features like this pond look beautiful and serene, but they will also encourage wildlife to visit.

Don't be too tidy

Within a garden, it is helpful for wildlife if you're not overly tidy. For example, instead of cutting herbaceous perennials down immediately after they have finished flowering, allow them to go to seed, so that birds can feed on them. The debris from plants can also be used for nesting material, so don't worry too much about collecting fallen leaves.

Many insects will also enjoy the hollow stems of herbaceous perennials, so consider leaving piles of them in the garden. At the edge of the garden, ideally in dappled shade, piles of logs and branches can be left to attract wood-dwelling invertebrates. These in turn will attract birds, hedgehogs, badgers etc., which will feed on them.

Wildlife-friendly lawns

Lawns will attract wildlife, particularly if some areas are left to grow long and allowed to flower and go to seed. If you have room, consider creating a wildflower meadow. Not only will it look beautiful but also lots of birds, butterflies, bees and other creatures will be attracted to the diverse habitat.

Provide a water source

All creatures will require a source of water, so consider building a wildlife pond if there is room. It should ideally have a gentle sloping edge so that creatures can access the pond for a drink without risk of falling in, and can also exit safely. If there isn't space for a pond, then a dish of water will offer passing flying insects such as honeybees and butterflies an opportunity to drink. Put stones in the saucer so the insects can perch on something.

Create a bug hotel

Providing a small home for a wide range of insects will increase biodiversity in the garden. It will attract pollinators, which will increase your yields of fruit and vegetables. It will also attract ladybirds, earwigs and lacewings, which will help reduce aphid infestations. And if you're lucky, you may even have hedgehogs and toads making a home, which will feed on slugs and snails.

You will need:

Old bricks

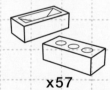

x57

Pieces of /timber

x4

Drill with various-size drill bits 3–8mm (⅛–⅓in) wide

Roof: roof tiles, slates or planks of wood

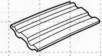

Pallets

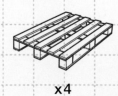

x4

Fillings: old flowerpots, straw, corrugated cardboard, dead/rotting wood, leaves, stones, woodchip, pinecones, grass, canes

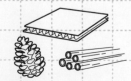

1 Identify a quiet area of the garden to place the bug hotel, where wildlife won't be disturbed. Ideally, it should be a level area that receives some sunlight during the day for bees, although shadier areas will attract toads and hedgehogs.

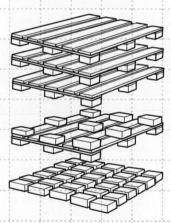

2 Make a base from bricks, leaving gaps between them to allow creatures to crawl among the spaces. Then lay four pallets on top of one another over the brick base, using more bricks to separate them, ensuring the stack is sturdy.

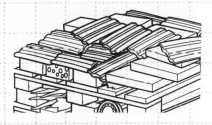

3 Drill holes in pieces of wood 3–8mm (⅛–⅓in) wide and about 12cm (5in) deep. These will attract solitary bees. Put these pieces of timber on the sunny side of the stack. Fill other sections of the bug hotel with rotting timber and old bark for beetles, spiders and woodlice.

4 Fill old flowerpots with straw, corrugated cardboard, deadwood or bundles of cut bamboo, and slot them into some of the gaps between the pallets. Stuff more leaves, straw, stones, woodchip, pinecones and grass in other bare areas to encourage wildlife to nest.

5 Finally, create a roof on top of the stack to try and keep everything – including the wildlife – dry. It doesn't have to be too fancy or perfect. Just a few roof tiles, slates or planks of wood over the top of the structure should help keep out the worst of the weather.

Make a lawn

There are many benefits to having a lawn, even if it is only a small one. The soft texture of grass makes it comfortable to sit, play or picnic on compared to harder decking or patios. Another bonus is that the lush verdant colour provides an attractive backdrop to herbaceous borders, and the absorption qualities reduce the risk of flooding.

Lawns are relatively inexpensive and easy to install, and if you allow the grass to grow, it will attract a range of wildlife into the garden. Commonly-found plants in a lawn, such as plantain, achillea, dandelions and annual meadow grass, will provide shelter, moisture, flowers and seed heads, attracting beetles, caterpillars, moths and butterflies, which in turn will encourage bats, birds and hedgehogs to come and forage.

Obviously, long grass isn't so comfortable to sit on or as practical to play on compared to nicely mown lawns. A good compromise is to leave specific areas of the lawn uncut for wildlife for the entire year and intersperse it with shorter cut areas for walking or sitting on. Areas of long grass only need to be 1–2 sq m (11–22 sq ft) for wildlife to benefit, although more is better.

Reduce mowing frequency

With regards to the shorter areas of lawn, raise the cutting blades to their highest setting and cut only every two or three weeks when necessary. If possible, avoid cutting the lawn from late summer, particularly if you're not going to be using the garden in winter, and restart the mowing regime in spring.

To maintain a green garden, consider using a push mower to cut small areas of lawn, as opposed to using petrol-driven machinery. Battery-powered lawn equipment is a good compromise if manually pushing a mower is too much physical work. Other equipment you may need is a strimmer, for keeping the edging of the lawn tidy and keeping long grass from overcrowding young trees or shrubs. For small spaces, a pair of hand shears or edging shears will do the job.

Long areas of grass should be cut just once a year, in late summer with hand shears. Leave the piles of cuttings by the side of the grass for a few days to allow any insects and tiny creatures to escape back into the garden. Then add the cuttings to the compost heap.

Right: Have a natural, attractive lawn by cutting it with hand shears – if it's a very small area – or push mowers instead of machinery, and by growing grass longer and allowing wildflowers to grow.

7

Create a pet-friendly garden

Pets enjoy being in the garden just as much as, or possibly even more than, us humans. There are many benefits for pets being able to roam outside, but it is important to ensure that the garden is a safe and secure environment for them, and that your garden itself is protected from a sometimes over-exuberant passion for digging up plants.

Let them dig!

Many pets love to dig in flowerbeds and lawns. If you have space, it is worth cordoning off safe areas for them to indulge in this passion of digging, to prevent them damaging other areas of the garden.

Hard-scape around plants

Putting gravel down around your plants and in your flowerbeds will help prevent digging, thereby adding a level of protection, and will make it easier for you to spot and clean up any unwanted toilet mess.

Plants that can harm pets

More comprehensive lists can be found online, but some of the more common plants to avoid due to their toxicity to pets are: aconite, astrantia, buttercup, chrysanthemum, daffodil, daphne, delphinium, foxglove, grape hydrangea, lily of the valley, laburnum, onion (including onion-family chives, leeks, garlic, peony, spring onion etc.), tomato, tulip, wisteria and yew.

| Aconite | Astrantia | Daphne | Lily of the valley | Laburnum | Wisteria |

Provide shade

Plant trees or shrubs to create shade for hot pets in the summer, if there are no natural shady areas in the garden. Providing a balance of light and shade will help your pet enjoy the garden.

Select the right plants

The good news is that many plants aren't toxic to pets. It is worth selecting robust non-toxic plants that will tolerate being knocked and brushed against.

Identify and limit hazards

Keep pets away from recently sown grass seed or wild meadow mixes, as awkward-shaped seeds can get lodged and become an irritant.

It is best to keep ponds fenced off from pets, thereby keeping other wildlife safe, as well as your pet.

Build barriers

Many pets have an inbuilt, or natural, awareness of plants that will make them ill and will avoid them. However, if you want to make sure your pet is protected from potentially harmful plants, and vice versa, it may be necessary to practise some 'petscaping' and create barriers or fences around flowerbeds or clusters of bulbs. Hedges and shrubs will provide a natural barrier, particularly for pets that can jump; while walls of natural, local stone can be attractive, inexpensive and effective.

Another method is to grow plants in raised beds, keeping them out of the reach of some pets, although those such as cats are agile, and excellent climbers, so will still be able to reach the plants.

Adorn shade with low-light plants

Gardens in shade present unique opportunities to gardeners, allowing them to grow stunning shade dwellers that thrive in these cooler, darker conditions. Add relevant shade dwellers to enhance the typical shady areas of a garden, or to maximize your shady growing spot.

Seating areas

Cool areas of the garden can be used to create relaxing seating areas, out of the heat of the direct sun on hot days. There can either be an entire seating area for alfresco dining, or just a bench to perch on to enjoy a cup of tea or a glass of wine. Surround seating areas with luxuriant shade-loving plants such as hardy male ferns (*Dryopteris affinis*), large-leaved hostas and colourful hellebores.

Pathways

Pathways are a useful method of dividing up a tricky shade area under trees. Create winding, curvy paths to add a touch of informality. Even in the tiniest of shady gardens, a natural path is a useful design feature that creates a sense of space and direction, emulating a woodland trail, just on a much smaller scale. Use natural materials such as bark or wood chippings for the surface,

and lengths of branches or logs for the sides. Edge with tall foxgloves (*Digitalis purpurea*), or, for something smaller, lady's mantle (*Alchemilla mollis*), with large sprays of yellow flowers, and spring-flowering bulbs such as daffodils, crocus, snowdrops and bluebells.

Ground cover

Low-maintenance, ground-cover woodland plants thrive in shade more than a lawn will – and are more beautiful. Suitable ground cover for shade includes the mat-forming evergreen Japanese spurge (*Pachysandra terminalis*) and the clump-forming Siberian bugloss (*Brunnera macrophylla*), which closely resembles forget-me-nots.

Water features

Creating a water feature adds a touch of tranquillity to a shady area. If there is a nearby power source, a water pump can be used to create the sound of running water, imitating a running stream, fountain or small waterfall. Alternatively, a still woodland-type of pool is an equally beautiful feature. Plants that will enjoy the moist, shady conditions around a pond or water feature include the broad-leaved false spikenard (*Maianthemum racemosum*) and foam flower (*Tiarella cordifolia*), with creamy white blooms.

Right: Create a rich tapestry of texture and a variety of hues in shady corners of the garden using a range of foliage plants such as hostas, dwarf pines, ferns and heucheras.

SHADE-LOVING PLANT SELECTOR

There are just as many exciting plants that thrive in shade as there are ones for direct sun. Many of these dwellers of the dark corners in the garden are bold and beautiful, providing impressive foliage and flowers. Many can be used as ground cover, imitating the conditions of a woodland garden, whereas others are suitable for climbing up fences or into tree canopies. Shade-loving trees and shrubs provide privacy and seclusion, while encouraging wildlife.

HOSTA

These popular herbaceous perennials are grown for their impressive large leaves, with colours including dusky blue, acid yellow, lime green and a few variegated types. Keep an eye on them, though, as slugs love them.

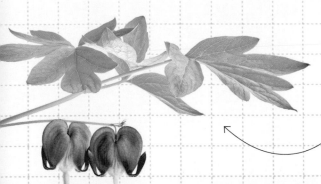

BLEEDING HEARTS
Lamprocapnos spectabilis

So called because they produce heart-shaped flowers with what looks like a droplet at the bottom. These herbaceous perennials prefer moist, cool conditions in partial shade, although they will tolerate some sunshine if the soil is kept damp.

JAPANESE MAPLE
Acer palmatum

These small deciduous trees with gorgeous-shaped palmate (five-lobed) leaves thrive in dappled shade. Many acers have attractive coloured foliage, including red, yellow and purple, and also have impressive autumn colour. Some have finely dissected leaves. They require protection from wind.

IVY
Hedera helix

Climbing ivy will cope with the deepest of shade and is tough as a pair of gardening boots. There are lots of varieties to choose from, including cream, silver and gold variegated forms. The flowers that appear in late autumn on this evergreen climber make it an important source of nectar for wildlife.

CLIMBING HYDRANGEA
Hydrangea petiolaris

Ideal for training up a north-facing wall or up a tree trunk, this climbing hydrangea shrub is native to the woodlands of Japan and the Korean peninsula. It produces lace-like white flower caps in summer, and its foliage turns a buttery yellow in autumn.

PERIWINKLE
Vinca minor

One of the best low-growing ground-cover plants for shade, periwinkle thrives in most soil conditions, producing masses of pinkish purple flowers. It is an evergreen perennial that will quickly cover bare soil, but it may need regular cutting back if you don't want it to take over other areas of the garden.

Create a dry garden

Dry gardens, sometimes known as gravel gardens, are a low-maintenance design solution for well-drained, arid soil conditions usually in full sun and with low rainfall. This style replicates similar growing conditions to some areas of the Mediterranean, with lots of interesting, drought-resistant plants thriving in the sizzling hot sun.

Gravel gardens use drought-resistant plants to survive dry, hot conditions. Often, the most suitable plants are small, silver-leaved plants such as lavender, artemisia, thyme, pervoskia and eryngium (sea holly). Ornamental grasses such as stipa, miscanthus, pennisetum and carex are also suitable, adding texture, drama and movement to the arid space. Apart from looking beautiful, another added benefit of this style of gardening is that it is reasonably low maintenance.

This style of garden requires well-drained soil in full sun. It probably won't work in heavy clay soil or in partial sun, although it may be possible to construct raised beds filled with light, free-draining soil and covered in gravel. It may be necessary to trim back any overhanging canopies and foliage to create more light.

Gravel gardens often don't have distinguishable pathways made of different materials to the flowerbeds. Instead the entire area, both beds and intended walkways or paths, are covered in gravel.

Plants are almost randomly or naturally interspersed throughout the gravel, with the paths becoming the gaps between them. This gives a natural feel and flow to the overall design, as though the plants are growing in their natural habitat.

Getting started

To create a dry garden, lightly dig over the intended planting area to break up any hardpans beneath the surface that may impede the drainage of surplus water. Incorporate a small amount of garden manure, but not too much, because this will encourage luxuriant growth, which will require more watering.

Soak the plants in a bucket for a couple of hours before planting. This will reduce the need for watering them once they are in the soil. Ideally, choose small, young plants, as they will establish in the dry garden better and will be less dependent on watering. Plants should be placed as randomly as possible to replicate nature, as if they have seeded

themselves into their positions. Once planted, cover the entire area with a 3–5cm- (1–2in-) deep layer of gravel and rake it level.

General maintenance

Allow the plants to seed naturally into other areas of the gravel. Watering should be kept to a minimum, if at all. After a couple of years, some of the plants may benefit from being dug up, divided into smaller pieces and replanted. This will keep the plants young and healthy. The layer of gravel should suppress most weeds, but occasionally pulling out a few rogue ones will be necessary.

Above left: Tall, eye-catching flowering plants such as sea holly (*Eringium*) add interest to swathes of grasses.

Above: Many plants with aromatic foliage such as lavender and salvias thrive in hot and arid conditions.

DRY GARDEN PLANT SELECTOR

Suitable plants for a dry garden are ones that will thrive in full sun, requiring minimal watering. They will flourish in arid conditions. Choose a mix of plants for their dramatic contrasts in size, shape and texture, and place them close to one another to exaggerate their architectural qualities.

WORMWOOD
Artemisia absinthium

Said by some to be the reason for the legendary hallucinogenic effect of the alcoholic spirit absinthe, wormwood is an upright perennial with silvery grey, finely cut foliage. It thrives in dry, arid conditions and is also suitable for herb gardens.

MEXICAN FLEABLANE
Erigeron karvinskianus

At first glance you might mistake this plant for the typical daisy you find in the lawn, and although the flower itself is similar, this plant is a clump-forming perennial with silvery green foliage that adores dry areas such as wall crevices and in among rocks and gravel.

AFRICAN LILIES
Agapanthus

Agapanthus have large drumstick-shaped flowers, usually in shades of blue, although there are also white and lilac varieties. These perennials have strap-like leaves and the seed heads remain bold and upright well into winter.

ELEPHANT GRASS
Miscanthus

If you want drama for prolonged periods of time, then this ornamental grass is ideal. It produces masses of arching foliage that can take on reddish tinges in winter, while impressive plumes of flowers appear in summer and last throughout winter.

SEA HOLLY
Eryngium

This dramatic perennial comes in many shapes and sizes depending on the species, but most of them have intensely dramatic, electric blue, silvery foliage. They have architectural, thistle-like flower heads with spiny stems.

VERBENA
Verbena bonariensis

A tall herbaceous perennial with masses of airy, purple foliage held aloft above the ground. They seed themselves prolifically around the garden, which adds to the natural effect of a gravel garden. Bees and butterflies love them, as they are rich in nectar and pollen.

Create a garden jungle

Lush exotic foliage and flowers can create a dramatic wow factor in a subtropical style of garden. Some 'jungle' plants can be large, so they often have even more impact when grown in small spaces. Many of the exotic-looking plants are surprisingly hardy and will survive outdoors.

Subtropical plants will ideally have rich, fertile soil in full sun. They will require watering during hot weather, emulating the heavy downpours found in tropical areas, which will enable the plants to produce their lush foliage.

For the full jungle effect, you'll need to feel completely immersed among plants. Create deep, wide borders offering layers of foliage, with the smallest plants at the front and larger ones at the back. However, drift some of the taller plants forward to create the effect of being among towering plants. This effect is easier in a smaller garden as you are closer to the plants, and the bigger plants feel as though they are towering above you.

Paths and seating

Paths made of timber decking add to the jungle theme. Even in a small garden, design paths so they lead out of sight, to give a sense of adventure and making a space feel bigger than it is. Knock in posts at intervals on either side of the path and link them with rope swags as handrails.

Tuck small seating areas in among the lush plants on areas of decking. Hammocks hung between two posts or trees also contribute to the atmosphere.

Get the look

For impressive foliage, try growing banana plants. The hardiest variety is the Japanese banana (*Musa basjoo*); it provides lush, tropical-looking foliage (although it won't reward you with any actual fruits). Cannas and ginger lilies will provide a similar effect. Alternatively, try cordylines, palms and tree ferns.

Dahlias, with their brightly coloured flowers, also add to a subtropical theme, while bamboo plants provide an excellent backdrop to a jungle-styled garden.

Clockwise from left: Ginger lilies, with their exotic flowers and lush foliage, are a wonderful addition to a tropical border.

Tall cannas are ideal for the back of a sunny flowerbed.

Create twisting paths packed with plants jostling next to each other for a true jungle effect.

11

Make friends with weeds

With a bit of judicious management and careful selection, you can keep on top of those plants in the garden that like to dominate or take over their environment. Or you can just learn to live with them, enjoy them for what they are and even reap some of their benefits.

A weed is simply a plant making a nuisance of itself by growing in the wrong place and interfering with the intended, cultivated plants, and they can range in size from a small dandelion to an enormous tree. One inescapable fact in gardening is that weeds will always find their way into a garden. The trick is to control or manage them efficiently so that their benefits can be enjoyed, rather than them becoming a problem.

What's the problem?

If weeds are left to grow among other plants, they will compete for nutrients in the soil and therefore deplete the availability of fertility for the existing plants. This can lead to nutrient deficiency in vegetable crops, fruit plants or ornamental plants.

In addition, weeds smother out the available light for other, cultivated plants, making the latter grow into a spindly or weakened shape, or killing them completely. For example, a patch of brambles would quickly outcompete a row of cabbages, probably resulting in the vegetables dying.

Benefits of weeds

There are, however, a number of benefits to these much-maligned plants. Weeds contribute to a balanced ecosystem in the garden and will attract many insect pollinators. For example, ivy and dandelions are a great source of nectar in early winter and early spring respectively, when there are few other plants available. During the growing season, the same insects will revisit your fruit and vegetable plants, pollinate them, which will increase yields.

Some weeds can also be used medicinally in the garden. Plants such as nettle and comfrey can be made into a rich liquid fertilizer (see page 140). Some weeds can be eaten or even made into a delicious wine or cordial.

If possible, allow some space for weeds to grow in the garden, to encourage biodiversity. The more variety of plants in your garden, the healthier your garden will be. And, let's face it, weeds are free, so it is best to make the most of them. Create a small area of the garden specifically for them, whether it is letting an area of lawn grow longer or dedicating a flowerbed to them.

Clockwise from top left: Bindweed is a very common and persistent weed but can be controlled by regularly pulling out its sprawling stems.

Flowers such as this clover are considered weeds by some, but bees love them for their nectar-rich flowers.

Ragwort is poisonous – to livestock particularly, which will graze on it – and it is advisable not to let it go to seed.

Keeping weeds under control

Although there are benefits to weeds, it is important they don't take over completely, and some control of their tendency to spread around the garden is needed. Otherwise, they will take over and existing plants may become overwhelmed. In addition, too many weeds can look untidy and messy in the garden.

Cover up – weeds will quickly colonize bare soil, and the simplest method of preventing them doing this is to cover the ground over. This blocks out light and therefore prevents seeds germinating. For small areas, you can use suppressing membrane, sometimes called landscape fabric, or even old carpet, if there aren't any chemicals in it that might leach into the soil. However, this isn't always practical in large areas, and it isn't good for the environment and biodiversity to retain large amounts of bare soil under synthetic materials.

Natural mulching materials – a preferred method, which is gentler on the environment, is to use old cardboard boxes as cover, as the material eventually breaks down into the soil, providing a source of carbon for the intended garden plants to grow in.

Covering bare soil with natural material such as rotted horse manure, garden compost or even slate, gravel and pebbles will suppress weeds for a while, although they will eventually start to grow through.

Get planting – without doubt, the most effective and environmentally sound way of covering over the soil to prevent weeds germinating is to grow garden plants in those areas. Plant your favourite plants into flowerbeds with bare ground and they will smother out the light and help prevent weeds from germinating and spreading.

Weed removal

Always allow room for a few weeds in your garden. If you must remove some from areas of the bed where they are having a detrimental effect on the plants, there are a couple of methods for removing them, depending on their type: annuals or perennials.

Annual weeds – will grow, flower, set seed and die all in the space of one year. They usually have a small root system but are prolific at seeding into new areas of the garden. Examples include fat hen, groundsel, hairy bittercress, annual meadow grass and chickweed.

The most effective method of removing annual weeds is to use a long-handled hoe. Push the hoe just below the surface of the soil, cutting through the roots of the annual weed. Gather the plants and add them to your compost heap.

Perennial weeds – live from year to year and have a pernicious set of roots that need to be removed entirely when weeding, to prevent the weed from continuing to grow. Examples include bindweed, knotweed, dandelion, stinging nettle, ground elder, couch grass and bramble.

It is important to remove all the root when weeding perennial weeds. Use a fork to dig up the root, taking care. Don't add perennial weeds to the compost heap as the roots will re-germinate.

Opposite: The best method of preventing weeds is to cover the ground with plants, such as with this hosta, where its large foliage blocks out the light below.

Above left: Hoeing is an efficient method of removing annual weeds, but be careful not to damage the roots of existing plants.

Above right: Some plants, such as this dandelion, have long tap roots. The entire root needs to be removed to ensure it does not regrow.

Grow scented plants that come to life at night

One of the pleasures of sitting outside in the garden on a warm summer's evening is the plethora of fragrances that are released from night-scented plants. Many of these plants attract night pollinators such as moths, too.

Carefully planning where to place your evening scented plants will enhance the magical experience of spending a summer's evening in the garden. Most scented plants will be more fragrant if they have had a chance to warm up in the afternoon sun, so try to place them in the garden where they will maximize the last rays of sunshine in the day. This is usually a south or southwest aspect.

Seating areas

Place evening plants near seating areas to intensify the heady, balmy aromas on a late summer's evening. To maximise the individual aroma of each plant, in a small garden it's best to grow just one or two plants with your favourite aromas. In larger spaces, night-perfumed plants can be placed in different areas so that the specific fragrances are encountered as you move around.

Opposite: Train scented climbers such as wisteria (shown), honeysuckle and star jasmine on walls to surround you with fragrance.

Edge pathways

Place these plants at the sides of garden paths to capture their beautiful aromas as you pass by. Many of the plants also have enchanting pale white flowers or foliage so that they can be seen in the moonlight, naturally highlighting which way to go.

These plants will encourage and entice you out into the night garden to explore or relax. After all, many of us are at work during the daylight hours and the garden is in darkness when we are back home, so by creating a scented garden that comes alive at night, you are extending the sensory pleasure it can provide.

Containers by the backdoor

Plant some fragrant plants in pots by the backdoor. On a warm evening with the door open, the fragrance will waft inside, filling the house with wonderful scents, a bit like a natural air freshener. Also place the plants below or outside bedroom windows, so that they can be smelt at night when lying in bed on a balmy evening with the windows open.

EVENING-SCENTED PLANT SELECTOR

It is surprising how many plants there are that release an enchanting fragrant aroma in the evening, intended to attract pollinating, night-flying moths, but inadvertently also enticing us into the garden. Evening-scented plants come in all shapes and sizes, including small herbaceous perennials, large shrubs and even climbers to cover pergolas and surround seating areas.

EVENING PRIMROSE
Oenothera biennis

This tall perennial opens its yellow flowers in early summer on warm evenings and releases a citrussy aroma. It is suitable for flowerbeds and borders, or can be grown in containers. It likes to seed around the garden so keep an eye on it, if you don't want it taking over.

HOLLY-LEAVED SWEET SPIRE
Itea ilicifolia

An attractive glossy-leaved evergreen shrub with a slightly lax habit, so benefits from being grown against a wall or fence, although it can be grown freestanding, too. It produces long catkins in summer with a honey aroma that pollinating moths find irresistible.

JAPANESE WISTERIA
Wisteria floribunda

Most wisteria are scented, but the floribunda is said to have the best aroma of all. Long racemes of highly fragrant blue flowers hang down from its branches. It is a climber so will need a structure to train it. Perfect for growing on a pergola over a seating area.

ANGEL'S TRUMPETS
Brugmansia

A slightly tender shrub originating from South America, but will survive outdoors in sheltered, frost-free areas. Alternatively, it can be grown in pots and taken indoors for winter. The shrub features large trumpet-shaped fragrant flowers up to 30cm (12in) long in a range of colours.

WOODLAND TOBACCO PLANT
Nicotiana sylvestris

This short-lived perennial produces masses of small, white, trumpet shaped flowers. It has a sweet aroma during the day but is even more intense at night. It is fairly tall so is suitable for the back of a border, or can be grown in containers. It seeds profusely around the garden.

DAME'S VIOLET
Hesperis matronalis

If you love the smell of violets, you will appreciate this intensely fragrant biennial or short-lived perennial. Dame's violet produces pink, mauve and white flowers with an aroma that intensifies during the evening. Perfect for rockeries, gravel gardens and cottage gardens.

Clockwise from left: The callicarpa shrub produces stunning, decorative, metallic-purple berries that can last until Christmas time.

Daphne is one of the most popular scented shrubs for a winter garden.

Create an impressive colour contrast in winter with the brilliant red stems of dogwood against the pure white trunk of birch trees.

13

Enjoy your garden in winter

Perhaps surprisingly, winter can be one of the best times of year to enjoy the garden. There are plants that perfume the cool air with their sweet fragrances, brilliantly coloured winter stems and trunks, and bright winter berries. In addition, there are lots of plants in flower.

Winter stems and trunks

Once deciduous trees and shrubs are denuded of their foliage in winter, some plants come into their own, as they are able to show off their spectacular bare winter stems. The most popular winter stems are willow and dogwood (*Cornus*), which come in a range of colours such as fiery reds, oranges and yellows, as well as deeper colours such as blacks and purples. They are often cut down to near ground level in early spring, once their winter display has been enjoyed, which will encourage fresh stems to be enjoyed the following year.

Even if it is too cold to go outside, if you plant just a single tree with an attractive trunk within sight of the kitchen window it is enough to brighten your day. There are a number of trees that have beautiful trunks, which are shown off to their best in winter. Birch are probably one of the most popular winter trees, with their dazzling white trunks, but there are also snake-bark maples (acers), with their stripy trunks, or the peeling bark of the paperbark maple (*Acer griseum*). Another attractive winter tree is the Tibetan cherry (*Prunus serrula*), with its deep-red ornamental trunk.

Scented shrubs

The cold air accentuates the fragrances from some of the early flowering trees and shrubs. Plant them near your front door so that you smell them each time you go in and out of the house. Some of the best shrubs for winter fragrance are dwarf sweet box (*Sarcococca hookeriana* var. *humilis*), daphne 'Jacqueline Postill' (*Daphne bholua*) and wintersweet (*Chimonanthus praecox*).

Flowers and berries

There are lots of brightly coloured flowers available in winter that look great in containers or in hanging baskets. Some of the most popular ones are hellebore, winter viola, heather and cyclamen. Snowdrops also start to emerge in late winter and make a great feature at the base of trees or a hedge.

Brightly coloured berries stand out in winter, particularly against the backdrop of evergreen foliage. Holly is the traditional plant grown for winter berries, but there are other great shrubs with equally impressive displays, such as skimmia, pyracantha (firethorn) and cranberry.

Create a winter container display

Brighten up the darkest of winter days with this colourful winter display.
Place it in full view of one of your windows so that you can admire it from
the warmth of your house, or by the front door so you can enjoy it every time
you leave or enter in the home. All of these plants are found at garden centres,
and are easy to maintain.

You will need:

Frost-resistant container

x1

Skimmia 'Rubella',
as central feature

x1

Broken crockery
('crocks') or stones

Bricks

x2 minimum

General-purpose peat-free
compost

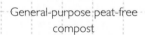

Selection of winter-
flowering plants, e.g. winter
pansies, viola, primula,
heather, hardy cyclamen

1. Choose a suitable container that
will be able to withstand the cold
weather. Wood, terracotta or
earthenware are suitable materials,
just ensure they are frost resistant
and have drainage holes in the
bottom.

2. Place a few broken crocks or stones
at the bottom of the container. They
will add weight to lighter plastic
containers, prevent compost being
washed through the holes and will
improve drainage.

3. Fill the container halfway up with
compost. Position your plants in the
container, placing the larger skimmia
shrub in the centre and the other,
smaller plants around the outside.

4. Backfill the rest of the container
with the compost, using your
fingertips to firm it around the
plants. The compost should come
up to just below the height of
the container.

5. Move the container into its final
position. Place the container on
level bricks, as this will help with
drainage. Give the plants a light
water to help settle the compost
around the roots of the plants.

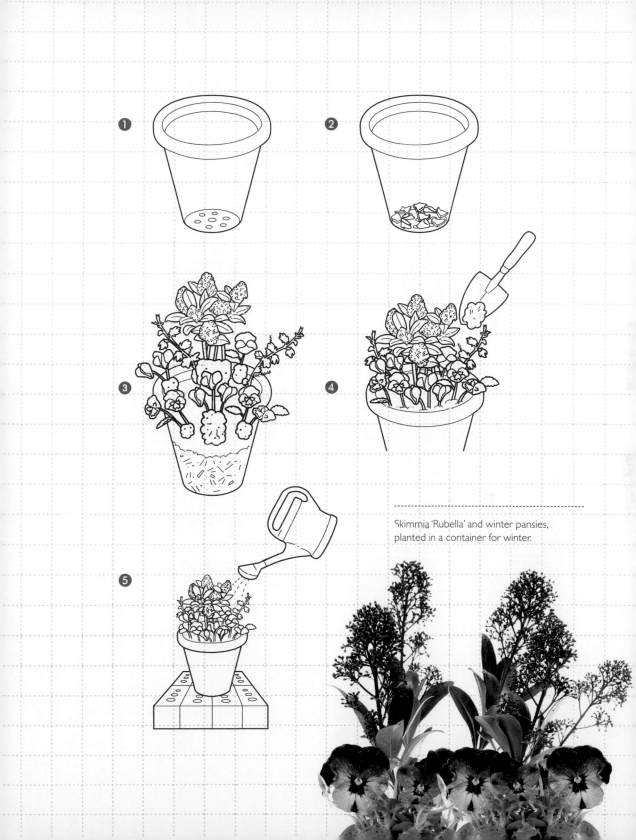

Skimmia 'Rubella' and winter pansies,
planted in a container for winter.

Elevate your garden

Sometimes in small gardens the only way is up, and this should be embraced, because it is fantastic to feel surrounded by plants at all levels, as if you are part of the garden. Bring an extra dimension to your garden by making the most of vertical surfaces, such as fences, walls and trellises.

Most gardens have walls or fences as boundaries, and these are ideal features to grow plants up. Climbing plants such as clematis, honeysuckle, Virginia creeper and wisteria can be trained up and along vertical structures using trellises or wires and vine eyes. Covering a vertical surface with plants makes a far more attractive, natural feature than a brown fence panel. In addition, it provides a habitat for birds and other wildlife.

Create intrigue

It's not just around the edges where vertical space can be utilized. Creating height inside the boundaries adds interest to an otherwise flat space. Fence panels, hedges or trellises can be erected strategically in areas inside the garden to train plants up.

As a design technique, using upright, vertical structures within the garden creates intrigue and the illusion of a larger garden, as the area can't be seen in its entirety at the same time. Having paths that lead behind a screen, hedge or trellis and out of sight is a clever trick for giving a sense of there being more garden behind.

Archways are a lovely way to create an entrance to a different area of the garden, leading the eye through into another area. Plants such as climbing roses or grapevines can be trained up and over them for attractive decor.

Keep things hidden

Creating height can also be used to hide unsightly features in the garden. Hedges, fences and trellises with plants trained over them can be used to hide compost heaps, dustbins or sheds.

Pergolas over seating areas is another great way of extending the growing area to overhead rather than losing a planting area due to tables and chairs.

Opposite left: Use fences and walls to train climber plants up, such as this black-eyed Susan vine (*Thunbergia alata*), for additional growing space.

Opposite right: Pergolas and archways are perfect for training climbing roses on, and you can smell their lovely fragrance as you walk past.

Plant a feature tree

Adding a feature tree is a simple way to create upright interest in the garden. Its canopy will provide shade in the summer and, depending on the type of tree, could produce seasons of interest, such as spring blossom, late summer fruit, autumn colour and attractive bark in winter.

15

Create a quiet nook

Gardens are the perfect place to escape from the day-to-day pressures of modern life. Relaxing in a quiet nook outside surrounded by beautiful plants and the sound of birdsong, and perhaps running water, is the perfect way to relax and enjoy the peacefulness of the garden. Even in the smallest of spaces it is possible to find your own slice of horticultural heaven.

One of the key elements to creating a quiet nook is giving the space a sense of seclusion, enclosure and privacy. This can be achieved with screening a seating area so that you are immersed in your immediate surroundings and separated from the other elements of the garden and house. There are various materials that can be used to surround the hidden nook, including fencing, trellises, weaved panels or stone walls.

A hedge sanctuary

Natural planting around a seating area also works well with either mixed informal hedging using plants, such as hawthorn, holly, elderflower and dog rose, or something more uniform and formal, such as a clipped yew hedge. Don't forget that deciduous plants lose their leaves in autumn, so that sense of seclusion may be lost for half of the year.

Another technique could be to create a gap in an existing, established thick hedge. By removing a few of the upright branches at the front of the

hedge with a saw or pair of loppers, a bench can be placed in the gap to create a quiet nook, surrounded by the greenery.

Willow arbour

A living willow arbour can be grown quickly around a bench in the garden to create a hidden space. Insert the living 'withies' (young willow shoots) about 5cm (2in) deep into the soil in a semicircle around the back of the seat, with 20cm (8in) between each one. In a year or two, the withies will have grown enough to bend their growing tips toward the centre above the bench. Simply tie them together to create a small willow arbour. In the summer, when the willows are in full leaf, they will form a dense canopy of foliage, creating a perfect hidden bolthole in the garden.

Block out noise

If you live near a busy road or have noisy neighbours, there are certain tricks you can use to soften their impact. One of these is to create a water feature with running water, such as a fountain or small waterfall. The idea is that the sound of the water will block out any irritating commotion outside of the garden. Even if it doesn't block out the noise entirely, the sound of water has a calming influence upon which you can focus instead.

For a more subtle, soothing atmosphere, try bamboo and ornamental grasses, as their rustling sounds when they sway gently in the wind are also relaxing.

Opposite: Tall green foliage plants such as ferns, hardy banana trees (shown) and bamboo will create a secluded hideaway.

Above left: Position benches in tucked away corners for privacy, but leave a discreet view to enjoy when sitting there.

Above right: Ornamental grasses such as this pampas grass grow quickly and become very tall, making them an ideal screening plant.

Get comfortable

If you plan on spending some time in your hidden nook – and you should! – then consider introducing some comfortable seating, with outdoor sofas, cushions and beanbags. Lounging in a hammock slung between two posts or trees is another lovely way to while away a few hours.

Think about the journey

If there is space, consider the pathways or access to the hidden nook. Sometimes it is nice to have a short, secluded journey to the end destination, as it adds the element of discovery to the hideaway. Make sure the hidden nook really is out of sight, using twisting paths or creating uprights such as archways or trellises leading up to it, to obscure it from the house.

Build a hideaway

Summerhouses create the perfect hideaways in the garden. Many can have all the creature comforts that you would expect in the house, such as electricity, log burners and Wi-Fi, although you may need to check with local authorities whether planning is required. If you are handy, then the most cost-effective way to create a summerhouse is to convert a shed. To create a magical atmosphere, it can be given a green roof with plants such as sempervivens and other succulents. The sides can be cladded with branches to replicate a log cabin in the woods. Outside, plant small trees such as birch, crab apple and cherry to continue the theme of a woodland hideaway, or grow some Japanese maples in pots.

If you're feeling adventurous, and have the space, consider a treehouse. A room with a view is the ultimate fashion accessory outside, providing you with a bird's-eye view of your garden and plants. Obviously, this does depend on having a suitably large and sturdy tree, although some specialist firms will also provide a support structure for a treehouse. If DIY isn't one of your strengths, then it is worth having an expert help with the construction.

Above left: Hammocks are perfect for whiling away some quiet, sleepy moments in the summer; simply tie them between two stout posts, trees or a hammock frame.

Above: Surround yourself with tall wildflowers and grass for full immersion in nature.

Right: For a bird's-eye view of the garden, a treehouse is a perfect garden retreat.

16

Design a no-fuss garden

Having a lovely garden to spend time in, full of wonderful plants, is something most of us desire but may not have much spare time to devote to achieving. Fortunately, there are many design ideas and lots of plants that require very little maintenance and care, while still creating a beautiful space outdoors.

Having a low-maintenance garden is not just about 'not having enough time'. There are other reasons for wanting to be less hands-on in the garden. These could include having a disability, not being fit and active enough, age, owning a rental or holiday property, living in rented accommodation or simply not being interested in gardening or plants and yet wanting to sit outside and enjoy the surroundings.

Elements of a no-fuss garden

Thankfully, having a low-maintenance garden doesn't have to involve simply paving over the entire garden. It is possible to be surrounded by lovely, beautiful plants for a minimal amount of effort, but it may be necessary to put in some work at the start to create the garden (although landscapers can be hired for this).

Gravel beds (see page 34) – use gravel as a mulch to suppress weeds and retain moisture. Most plants in a gravel bed are slow growing, requiring no staking, pruning or watering.

Slow-growing evergreen shrubs – are ideal for hedging or structure, as they usually require minimal pruning or trimming. A good example is yew. Clump-forming bamboos are also useful, as they rarely need pruning yet provide upright structure and screening.

Long-grass lawns – encourage wildlife; perhaps just mow a path through larger areas if access is needed.

Ornamental grasses, hardy succulents and some perennial herbs – usually require hardly any maintenance such as staking, pruning or cutting back.

Irrigation system – if you want to grow plants in containers, establish a watering system attached to a water butt or tap, to avoid having to water regularly. Try piercing small holes in an old hose pipe (drip hose) and attaching it to a garden tap with a timer to come on once a day for 15 minutes. Lay the hose among the containers, with holes placed by individual plants.

Clockwise from left: Self-seeding flowers such as forget-me-nots require little or no maintenance yet flower profusely, rewarding you with a bright, colourful display each spring.

Echeveria succulents hardly ever need watering.

Use slow-growing shrubs and plenty of ground cover to reduce the need for weeding and pruning.

Things to avoid

Here are some of the more time-consuming, demanding things that you may want to avoid if you're pushed for time:

Lawns, although you could cut them less frequently than the recommended once a week.

Bedding plants need planting each spring and removing each summer, and require watering.

Annual flowers need to be sown and planted each year, and many require staking.

Tender plants require moving indoors each winter or covering up with a fleece in cold weather.

Containers and hanging baskets need watering most days in summer, and need repotting or replanting most years.

Vegetables are mostly annuals and need bed preparation, sowing, planting out and harvesting.

Fruit trees and climbers require pruning and training each year.

Empty ground – weeds will quickly spread into it.

Greenhouse plants need watering, vents need to be opened and shading erected in hot weather.

No dig – create a no-dig garden (see page 146) for growing vegetables, if you want to grow your own food. This involves placing layers of cardboard and compost over the soil and growing the plants through them. Although vegetable growing still requires some work, there is far less weeding, raking and digging using this method than conventional gardening.

Rock gardens (see page 62) – can be created with alpine plants as they are slow growing, requiring little maintenance.

Small raised beds – create pockets of beds on the patio or decking and fill them with herbaceous perennials. Smaller beds of about 1m (3ft) across are easier to manage than larger ones, and being off the ground means there is far less bending over for planting, weeding and general maintenance.

Clockwise from top left: Perennial vegetables such as asparagus require less work than annuals if you do not have time to sow seeds and clear beds each year.

Creating a rockery might be hard work initially, but once established, alpines are slow growing, and the surrounding rocks and gravel suppress any weeds.

Choose the right plants for growing in gravel and you can relax and enjoy the stunning results – there will be very little weeding or watering required.

17

Make a rock garden

Re-creating a stunning mountain-top scene or the rugged landscape of a cliff face provides an intriguing challenge for home gardeners. Rock gardens packed with alpine plants can be created in the tiniest of gardens, and once constructed are low maintenance, with alpine plants creating seasonal interest for much of the year.

Many of the alpine-type plants are small, making them suitable for planting into crevices or spaces including troughs, sinks and containers. As there are lots of different alpines to choose from, it is possible to have plants of interest for every season.

For those on the move or renters, alpine containers can be picked up and taken to a new house every time you relocate. Due to the compact size of these plants, plenty of interest can be crammed into the tiniest of spaces.

Create a rockery

Building a rockery is all about trying to emulate similar conditions to those found on a mountain. This includes free-draining soil and lots of rocks, nooks and crevices for low-growing plants to shelter from the wind. It may be necessary to cut back overhanging branches to allow light into the area, as full sunlight is one of the prerequisites of most alpines. If there is heavy soil in the garden, rockeries can be made in raised beds filled with well-drained soil and horticultural grit.

Start placing the rocks or stones into position. It is usually easiest to start with a few of the larger rocks then place smaller ones around them. Create small nooks and crevices among the smaller rocks, which can then be used as planting holes. Dig out small hollows for the larger rocks to sit in, so they are secure. A crowbar can be used to lever the largest rocks into position.

Planting alpines

Make an alpine compost mix consisting of horticultural grit, leaf mould and loam in equal quantities. Remove the alpine plants from their pots, and plant them into the alpine compost mix. Use more horticultural grit to top dress the surface of the soil, to a depth of 2cm (1in) around the alpine plants as they do not like to get their leaves wet. This will also help prevent competing weeds from germinating.

Opposite: Use gravel and/or grit between rocks to prevent weeds germinating and to retain moisture for the alpine plants.

Sourcing stones

Stones and rocks can be purchased
from hard-landscaping suppliers.
Often, second-hand stones are
advertised in online marketplaces
for free. If possible, choose materials
that fit in with the local environment
and natural colours. Always look for
sustainable resources.

ALPINE PLANT SELECTOR

Alpine plants love well-drained soil in full sun. Their name originates from the Alps mountain range, where these plants were originally collected, although now the term refers to any hardy, low-growing plant that likes free-draining conditions. Alpine plants are usually small, as in their natural, exposed environment they hide out of the wind in crevices or behind rocks.

GROMWELL 'HEAVENLY BLUE'
Lithodora diffusa 'Heavenly Blue'

This low-growing evergreen shrub grows up to 40cm (16in) in height and width, and has vivid, dark blue flowers. Originating from Turkey and Greece, Lithodora means 'stone gift' in Greek, and the shrub is so named due to its love of rocky soils.

SISKIYOU LEWISIA
Lewisia cotyledon

A popular low-growing perennial that forms dark evergreen, fleshy, spoon-shaped foliage emerging from rosettes. The common form has pink funnel-shaped flowers, but there are hybrids with other colours including white, purple, orange, yellow and magenta.

BELLFLOWER 'BIRCH HYBRID'
Campanula 'Birch hybrid'

An evergreen perennial that produces masses of small, attractive, bell-shaped, violet-coloured flowers in summer, held aloft above a trailing mound of dark green leaves. Ideal for cascading over rocks or the sides of containers.

PINK 'POP STAR'
Dianthus 'Pop Star'

A popular alpine pink perennial with fringe-edged, lavender-pink double flowers with a deep cherry centre and an attractive scent. The flowers sit on a mound on foliage and are ideal for a rockery in full sun, in well-drained alkaline soil, or grown in containers.

SAXIFRAGE 'WHITEHILL'
Saxifraga 'Whitehill'

An attractive evergreen perennial grown for its rosettes of silvery-grey foliage, tinged red-purple at the base. It produces small starry white flowers on long arching stems from spring to mid-summer. Ideal for rockeries or containers filled with gritty compost.

COBWEB HOUSELEEK
Sempervivum arachnoideum

An evergreen succulent with green and reddish fleshy rosettes, with distinctive hairs that make it look as though the plant is covered in cobwebs. It produces attractive, pink, starry-shaped flowers in summer. Originating from the Alps, Apennines and Carpathian mountains it thrives in hot, dry conditions.

18

Introduce a water feature

A water feature can transform an outside space into an oasis for wildlife where they will be encouraged to drink or bathe. It can also create an atmosphere or mood within the garden, with flowing water such as a mini waterfall evoking vibrancy with its movement and sound, or a still pool inviting peacefulness, reflectiveness and tranquillity.

There are various ways water can be introduced into a garden, ranging from a wildlife pond to a tiny water fountain in a plant pot. Whatever your ambitions, water can be used as a focal point in the garden. Placing a water feature near the house will bring wildlife closer for those people who are less mobile, while positioning a pond further away down a path can create a destination distinct and separate from other areas of the garden.

In a small garden, it is best to choose a water feature style that fits in with the theme of the rest of the garden. A wildlife pond could look incongruous in a modern chic garden; similarly, a formal pond could appear too much of a contrast in a potager or cottage garden.

Practical considerations

There are lots of practical considerations with a pond, to make the most of this space. With regard to size, make the pond as large as possible, with a minimum depth of 50cm (1½ft) and surface area no smaller than 4.5sq m (48sq ft), otherwise there

will be problems with algae, excessive temperature variation, water evaporation and inadequate water oxygenation. The larger the surface area, the deeper the pond should be. If space is really tight, a mini water feature such as a container pond or wall fountain is a better option.

If a formal pond is being constructed with electricity supply (lighting/pump) it may be best to seek the help of a professional.

To avoid the water turning green with algae, it is best to place a pond in dappled shade, out of full sun. A build-up of debris such as falling leaves can increase the nutrients in the pond, so it may be necessary to scoop out leaves with a net in autumn.

Left: With the right containers you can set up a rustic cascading water feature, enhancing the tranquillity of your garden.

Create a small wildlife pond

A water feature is a key ingredient to creating interest and is a focal point in any garden. A wildlife pond is the simplest way to achieve this, enticing small mammals, birds and insects into the garden, and requiring no electricity for water pumps. A wildlife pond can be whatever size you want, and can be incorporated into the smallest of courtyard gardens.

You will need:

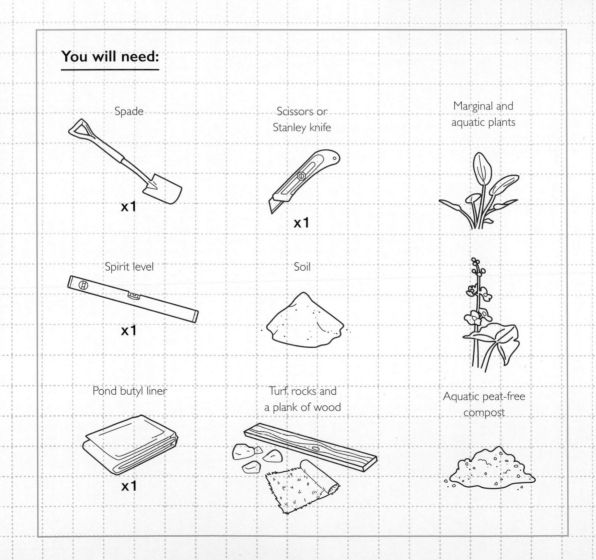

Spade
x1

Scissors or
Stanley knife
x1

Marginal and
aquatic plants

Spirit level
x1

Soil

Pond butyl liner
x1

Turf, rocks and
a plank of wood

Aquatic peat-free
compost

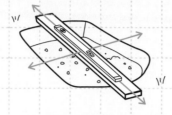

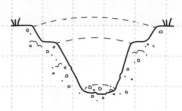

1 Identify the location of the pond, which should ideally be in dappled shade and in a quiet area where wildlife won't be disturbed. Outline the shape of the pond using sand or a hosepipe.

2 Dig out the pond. Use a spirit level and lay it on a length of wood, then place it across the entire hole of the pond, to ensure the levels are correct around the sides.

3 Ensure the pond has different depths, with shallow, sloping areas at the sides for wildlife to reach the water without falling in. It should have a 10cm- (4in-) wide, deep shelf around the outside for marginal plants and a deeper area of about 40cm (16in).

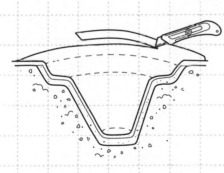

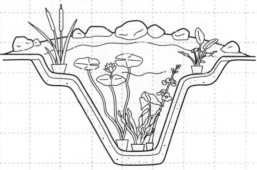

4 Remove any sharp stones or sharp objects that might puncture the pond liner. Then lay a base of 2cm (1in) of sand on the bottom and sides, and cover the area with a butyl pond liner, leaving an excess of 20cm (8in) around the perimeter edge, and cutting to size with scissors or a Stanley knife.

5 Cover the edges with soil, then lay turf, rocks and plants near or overhanging the edges of the pond, making it look as natural as possible. Plant marginal plants in aquatic compost along the shelves of the pond, and aquatic plants in the deeper areas.

Grow something unusual – a loofah

Loofahs (*Luffa cylindrica*) are fun to grow and have a practical use, too, because the fruit can be used as a sponge to clean windows, dishes or even yourself – much better for the environment than plastic-based sponges. Loofahs last for ages, and can even be put in a dishwasher or washing machine to clean them.

It's always good to challenge yourself, and growing a loofah will require a bit more work than other, simpler plants. Loofahs are a type of gourd, closely related to pumpkins, marrows and cucumbers. You will need a propagator to get them started, then a conservatory, greenhouse or sunny spot inside to cultivate one, as they require warmth to ripen fully.

Getting started

They need warmth to get them off to a good start, and a long growing season, so they should be germinated in a propagator in early spring. Soak the seeds in warm water first for a few hours, then sow each seed in a 9cm (3½in) pot filled with a general-purpose peat-free compost. Push each seed to about 2cm (¾in) below the level of the soil. Place it inside the propagator and keep it at a constant temperature of 25°C (77°F). Water the plant every few days when the compost feels dry.

Transferring and growing

After a few weeks, when they have reached about 6cm (2in) high, transfer the seedlings into individual larger pots, about the size of a bucket. Fill the pot with general-purpose peat-free compost and place the young seedling into it. Place it in a greenhouse or conservatory, or in a room in the house where it will receive sunlight.

Loofahs are climbers, so they will need a wigwam made from canes for their tendrils to grab on to. Allow each plant to form about three or four fruits, and after this, remove any more that appear.

Ready for picking

Loofahs are ready in late autumn, when the fruit has firmed up. Remove individual fruits from the plant and strip the exterior skin away to reveal the fibrous skeleton beneath, the part to keep as a loofah for cleaning. Remove the seeds. Leave the loofah to dry, then cut to the required size.

Right: Loofahs are climbing plants so will need a support system to help them grow upward.

Other unusual veg to grow:

- Purple carrots instead of traditional orange ones. You can also try round carrots, the size of golf balls; ideal for shallow window boxes.

- Oca, also known as New Zealand yam, as an alternative to potato.

- Romanesco – a quirky-looking cauliflower with a golden-green head, closely related to broccoli.

Make an easy-access veg garden

Raised beds can be any size, from a dustbin filled with soil to a few metres across. They offer a solution for those with back problems, as the bed's raised height saves bending over. Fill beds with lovely enriched compost to grow delicious vegetables, no matter the soil type or size of your garden.

Most raised beds are made from timber in a rectangle shape, although they can be any shape you desire. If DIY isn't your thing, kits can be bought that simply slot together and are placed in the garden. A more cost-effective option is to simply cut and screw lengths of exterior timber into a rectangle shape, using batons in each corner to hold the structure together.

Dimensions

Raised beds can be any size, but are usually no more than 2m wide, so that all areas of the bed can be reached and maintained without having to walk on the beds, which would compact the soil. If constructing a few raised beds, consider access between them, as it may be necessary to push a wheelbarrow between them.

There are no hard and fast rules regarding the height of a raised bed. Most vegetables are shallow-rooting, so 20cm (8in) depth should suffice. However, raising beds up to around 1m (3ft) high will save on back-breaking work when weeding, planting and harvesting.

Filling the beds

Fill raised beds with quality peat-free compost. Or, a cost-effective way of filling it is to treat it as a compost heap for the first year, adding fruit and vegetable peelings and mixing them with fallen leaves, cardboard and newspaper. Turned regularly, the soil will be ready for planting after a year, but still add a top layer of peat-free compost.

Mini raised 'beds' for potatoes

Hardly taking up any space, potatoes can be grown in tiny raised beds made from old compost bags or robust bin liners. In spring, roll down the sides of the bag to about a third of its height. Add a layer of general-purpose peat-free compost to the bottom of the bag, to a depth of 10cm (4in).

Place two seed potatoes on the compost and cover over with another 10cm (4in) of compost. In a few weeks, when emerging potato shoots start to appear, roll the bag up to halfway and top up with more compost. Continue adding compost and rolling up the bag as shoots emerge, until the bag is full of compost. Harvest the potatoes when the plants have finished flowering.

Clockwise from top: Raised beds are easy to make yourself from scrap timber, and make attractive, natural-looking features.

Experiment with container size and tuber spacing for potatoes to find the best combination.

Herbs thrive in most raised beds because they enjoy the free-draining conditions.

21

Harvest vegetables in pots

Most vegetables can be grown in containers, and although they might not yield as much as if growing in the ground, you can still expect a respectable harvest. One advantage to growing in pots is ease of care, as weeding is minimal. Also, pots can be moved around to follow the sun.

Most containers are suitable, just ensure they are frost-hardy if growing vegetables all year round, and have drainage holes. Vegetables in containers will have limited access to water and nutrients, so they should have a depth of at least 30cm (12in) to prevent them drying out too quickly.

Most plants prefer full sun, but there are a few that will cope with light shade, such as lettuce, chard, beetroot, broad beans, kale, cabbage and most salad-leaf crops.

Sowing and growing

The majority of vegetables are annuals, so you will need to sow them at the start of each year – little and often is the key to having a continual supply for much of the year. With careful planning, it is possible to successional sow (sow a crop every couple of weeks), in order to continue harvesting them in the same season, just a few weeks apart. This technique will keep a steady, manageable flow.

If you have access to soil from your garden, it can be mixed with rotted garden compost at a 1:1 ratio. Alternatively, soil-based container compost can be used, or general-purpose peat-free compost.

Veg for pots

Almost anything can be grown in a container if given enough food and feed, but root crops such as carrots and parsnips will need to be deep enough to be able to grow fully in length. However, there are round or globe carrots that are well worth trying in shallower pots. Plants requiring a heavier soil, such as the cabbage (brassica) family – brussels sprouts, kale and broccoli – need enough feed and space to thrive.

Try planting:

- Beetroot
- Carrots
- Broad beans
- Tomatoes
- Peas
- Rocket
- Lettuce
- Any cut-and-come-again salad leaves

Watering and feeding

In warm, dry weather, vegetable plants tend to need watering on a daily basis. Vegetables in containers will also benefit from a weekly liquid feed as they grow. A liquid feed rich in potassium, such as a tomato feed, is ideal.

Clockwise from top left: Repurpose car tyres as alternatives to containers; they keep the soil nice and warm.

When using containers to grow carrots, make sure they are deep enough to accommodate the roots.

Compact your leafy vegetables and salad leaves in portable troughs for ultimate convenience.

Most vegetables can be grown in almost any type of container, including a pair of old boots.

Grow vertical veggies

Many vegetables have climbing, scrambling or trailing habits, so if you are short of space on the ground, you can try growing crops upward and downward. Not only is this a great way to maximize space, but these vegetables also make beautiful upright features in their own right, with colourful flowers and crops at eye level. Another benefit is that they can be harvested at a comfortable height.

Suitable climbing plants include: runner beans, French beans, outdoor cucumbers, cordon tomatoes and cucamelons. All of these plants require full sun, and shouldn't be planted outside until the risk of frost has passed.

Wigwams

The easiest way to grow vertically is on a homemade wigwam-shaped structure. Wigwams are easy to construct, and, as well as being practical for training vegetables, they make a focal point in the garden. Simply push twelve 2.5m (8ft) hazel canes or bamboos into the ground in a 1m (3ft) circle, then pull them together near the top and tie with garden twine to secure. One or two climbing plants can be placed at the base of each vertical upright and trained upward.

Pallet wall

One of the simplest and cheapest ways to grow vertically is to use pallets. Turn a pallet onto its side and staple a landscape fabric to the back of it, or attach a sheet of marine ply with screws. Secure

the pallet to posts on each side of a fence or wall. Another pallet can be placed on this one to double the height, if made secure by attaching it to two sturdy vertical posts either side. Alternatively, pallets can be screwed to a wall or fence.

Remove about a third of the slats off the front of the pallets, retaining just a couple of slats near the middle and a couple near the base. Use some of the removed slats to make two shelves inside the pallet, one at the base and one halfway up. Fill the shelves with peat-free compost. Plant trailing gourds such as pumpkins, courgettes or squash plants with a space of about 40cm (16in) between each plant (two or three plants on each row). They can either be left to trail downward or trained upward over fences or even into trees or hedges.

Freestanding vertical containers

If you don't have any walls or fences to attach growing structures to, freestanding stackable containers that are built up from the ground can be used instead. These make attractive dividers and screens between different sections of the garden.

Right: Sweet peas and runner beans are the classic wigwam-grown plants – both are productive and attractive.

Below: Pallets are easy to find and can be modified into a vertical garden structure with effective planting pockets.

Buying vertical training systems

Wall-hanging planting modules can be purchased and fixed to fences and walls. Some have internal pockets or cavities that can be filled with water occasionally, reducing the need to water them. The pockets are filled with compost and almost any vegetables can be grown in them.

VEGETABLE PLANT SELECTOR

Growing your own vegetables at home is a rewarding pastime, and there are hundreds of varieties to choose from. You can choose veggies that don't take up much space, and, obviously, those you like the taste of. Also, you can select vegetables that are either not readily available in the shops or that are expensive.

ONIONS

Onions can be grown from seed, but the easiest way is from 'sets', which are basically tiny onions. Plant them in spring in full sun and in well-drained soil. Keep them watered and they will swell up to the size of a large onion by the end of the season.

BROAD BEANS

One of the hardiest members of the bean family, and usually one of the first vegetables to crop in spring. They can be sown in autumn, winter or spring, and produce delicious pods full of healthy beans perfect for casseroles, pasta dishes and stir-fries.

CARROTS

Preferring light, well-drained soils, carrots are usually sown thinly every few weeks in spring and summer for a regular supply. Thin out seedlings to leave one every 5cm (2in) once they reach 5cm (2in) high, to ensure the remaining ones are bigger.

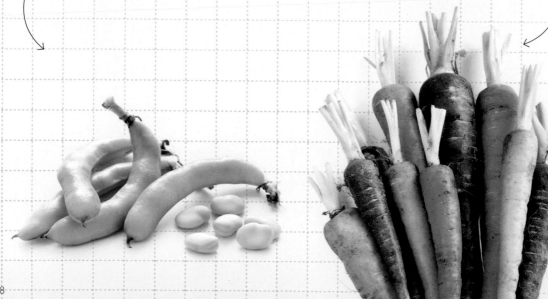

BEETROOT

Beetroot is tolerant of partial shade and a good choice for growing in a container. Sow directly in shallow drills, and thin out after a few weeks to leave the others to swell. Make regular sowings every few weeks for a plentiful supply.

SWEETCORN

Plant sweetcorn outside when the risk of frost has passed; they can be started off earlier indoors to give them a good start. Plant in a grid pattern instead of a single row, as they are wind pollinated.

PUMPKINS

Coming in all shapes and sizes, plant pumpkins outside when the risk of frost has passed. They like a rich soil, so add lots of compost. Some gardeners even grow pumpkins on top of a compost heap.

23

Garden on a tabletop

As the name suggests, tabletop gardening is simply growing plants, usually vegetables, on top of a table. The advantage of this system is the plants are at a comfortable height for both growing and harvesting – ideal for those of us with bad backs. Suitable tables can be home-made from scratch, or existing ones can be transformed to provide an edible feast.

Tabletop gardening is similar to growing vegetables in a raised bed, with the same benefits, such as saving on bending over, as the plants are at a comfortable height, and being able to add suitable compost if the garden has poor or no soil.

The difference is that raised beds don't usually have bottoms and are placed directly onto the soil, whereas tabletops are suspended on legs over a hard surface such as a patio. This means tables can be moved easily around the garden. In fact, some are placed on wheels or castors so they can be pushed into position.

Compost

Tabletops usually require less compost than a raised bed due to the space between the growing area and the ground. This makes them suitable for shallow-rooting vegetables such as lettuce, rocket and most other salad leaves. Radishes, baby beetroots, globe carrots and strawberries are also suitable.

Simply sow the seeds in a 1cm- (½in-) deep row (sometimes called a drill) and cover with peat-free compost every few weeks from early spring until late summer, and harvest them regularly. Place the tabletop near your backdoor, and you can pop outside to pick vegetables whenever required.

Pallet tabletop

The simplest way to construct your own tabletop garden is with pallets. Put two pallets on their sides to create the legs, or supports, at each end of the table. Lie a third pallet on top of the other two to create the tabletop itself. It helps to turn the top pallet upside down, so the majority of the slats are underneath, facing the ground. This will best support the compost above it and also makes a larger planting area. Line the underneath of the top pallet with old compost bags to stop the compost falling through the slats. Screw the pallets together to secure.

Push compost between the slats then plant or sow into the pallet. The advantage of the slats is that they prevent competing weeds popping up between the rows.

Make your own

Alternatively, take an old yet sturdy garden table and attach sides to it with 15cm- (6in-) wide timber all around the outside. Drill a few drainage holes in the tabletop. Simply fill the tabletop with peat-free compost and start sowing. Attach castor wheels to the legs so that it can be moved easily.

Opposite left: Bespoke tabletop planters can be purchased from most garden centres and are useful features.

Opposite right: It's easy to cobble together a tabletop planter using old timber.

Above: Old dressing tables and other furniture can be used as tabletop planters; treat the wood for outdoor use to make them last longer.

Grow a vine through a table

Create a touch of the Mediterranean in your back garden by dining under a grapevine canopy. Sitting under a grapevine will provide welcome shade on hot days, and the grapes will provide enough fruit to provide a few bottles of homemade wine. It will take two to three years for the grape vine to climb up the parasol fully, but once established it will make a beautiful feature on the patio.

You will need:

Wooden garden table with a hole in the centre

x1

Old parasol and base

x1

Outdoor grapevine such as 'Chardonnay' or 'Bacchus'

x1

Small, handheld jigsaw

x1

Spade

x1

Secateurs

x1

Garden twine

General-purpose peat-free compost

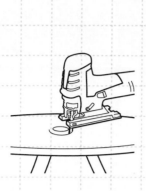

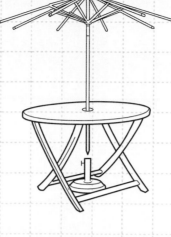

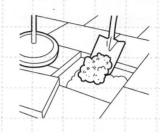

① Create a 10cm- (4in-) wide hole in the centre of the table using the jigsaw. Position the table where you will want to use it, as once in position it is difficult to move as the grapevine will be growing through it.

② Remove the fabric from the parasol. The framework of the parasol is going to be the training structure for the grapevine. Position the parasol base directly under the hole in the table. Push the parasol through the table and into the base.

③ Using a spade, remove a patio slab from next to the parasol base, remove any hardcore and rubble, then add compost to a depth of 30cm (12in). Plant the grapevine into the compost and firm it in before watering the roots.

④ Use secateurs to prune back all the side shoots but leave the top (the highest or leading shoot) unpruned. Tie this top shoot to the parasol using garden twine. Each year continue to train the top shoot until it reaches the canopy of the parasol.

⑤ When the grapevine reaches the framework of the canopy, allow some of the side shoots to grow and fill out to the shape of the original parasol. Each year cut back the new growth to two buds, leaving the original side shoots in place to retain the framework.

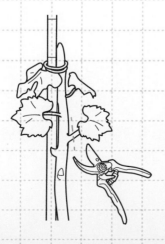

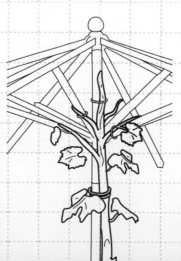

Delicious bunches of grapes will be ready to harvest in autumn and can be squeezed to make homemade wine or grape juice.

24

Train scented climbers around the garden

Scented climbing plants fill a garden with amazing fragrance. If there is a shortage of space, they can be trained easily on walls, fences or even the outside of the house. This means you can be surrounded with wall-to-wall aroma when spending time outdoors.

There are lots of scented climbing plants to choose from, and if selected carefully, it is possible to have fragrance all year round.

Rose swags

Roses are probably the most popular scented climbers, and they are often grown on 'swags', which are lengths of thick rope trained from one upright structure to another. They make beautiful features in a garden, provide a screen in summer when their foliage is out and are a great way of utilizing space if there are no walls or fences to train on.

Training systems

Some plants have tendrils and so are self-clinging and don't need any support – but be careful they don't damage any brickwork, as they can encroach into the pointing between the bricks. Other climbers will need a structure to scramble up, and will need tying in each year. Check the plant label before purchasing if you are unsure.

Trellis – a trellis can be attached to a fence or a wall, or can be an entire upright structure in its own right. It is useful for dividing up small areas of the garden. A trellis with a climber can be used to hide features such as a bin, and the fragrance from the plant is like having a natural air freshener nearby.

Archways – another great way of training scented plants; the aroma will drift down as you walk beneath the arch.

Porch and shed – if you have a porch, a scented rose can be trained over it, making an inviting feature when you come home. Not only does it look good, but it smells lovely, too. Sheds are also useful structures for training a climber, and a scented climbing plant is far more attractive than a felt roof. And there will be a lovely smell as you work away inside.

Top: Roses are the quintessential cottage-garden plant and make a beautifully fragrant feature in summer when rambling over an archway, swag or trellis.

Far right: This evergreen Japanese honeysuckle (*Lonicera japonica* 'Halliana') will keep its shiny leaves after flowering, offering year-round coverage.

Sweet peas

For a quick and inexpensive option, sow sweet peas in early spring and plant them out in late spring. Train them up an archway, trellis or wigwam made from bamboo. They will flower all summer long, enriching the garden with their sweet perfume.

SCENTED CLIMBING PLANT SELECTOR

There are numerous scented climbers that can be used to fragrance the garden. Most are either deciduous or evergreen perennials, but there are some fragrant annuals, too.

ARMAND CLEMATIS
Clematis armandii

This is a great climber suitable for shade or sun, with white flowers providing their sweet aroma in early spring. The plant originates from China and has attractive, thick evergreen leaves, so it can provide a privacy screen all year round.

ENGLISH CLIMBING ROSE
Rosa 'Gertrude Jekyll'

Named after the famous Arts and Crafts garden designer from the Edwardian period, this short climbing rose has an intoxicating floral fragrance. The plant is covered with bright pink double flowers for most of summer.

STAR JASMINE
Trachelospermum jasminoides

Ideal for a sunny, sheltered wall, although it will tolerate partial shade, this evergreen woody climber produces masses of star-shaped white flowers in summer. It is a fast grower, although in a cooler area its growth will be slower.

CHOCOLATE VINE
Akebia quinata

A semi-evergreen climber with unusual, cup-shaped, reddish-purple flowers during summer that have a chocolatey-vanilla fragrance. In warmer areas it retains its leaves in winter, but it loses them in colder regions. In warm summers it produces edible sausage-shaped fruits.

COMMON JASMINE
Jasminum officinale

A gorgeous, woody, evergreen climber producing white flowers with the sweetest of fragrances. It flowers all summer long but is quite rampant, so give it a large place to spread. Slightly tender, so train it on a sunny, warm wall for protection.

MORNING GLORY 'FRAGRANT SKY'
Ipomoea lindheimeri

This climber produces pretty, lobed foliage and lavender-blue, trumpet-shaped flowers that are highly fragrant. They are perfect for growing up small obelisks in a sunny border.

25

Grow a screen for shelter and privacy

Natural screens such as hedges are useful in a garden as they provide a sheltered space for both yourself and your plants. They can also offer screening from neighbours, while providing a wildlife haven for birds, insects and other small creatures.

Types of screen

There are a number of different screens that can be used in a garden. Hedges are ideal for creating privacy and shelter as they filter the wind and provide a habitat for wildlife. They also offer seasonal interest with blossom and berries. Trellis or fence panels can also be used as a screen, with plants trained up them for wildlife and to make a more attractive feature. Finally, you can weave your own panel (see page 90). These will last a few years and are free to make if you can find the appropriate material (usually hazel or willow), and look fantastic.

What type of hedge?

If you choose a hedge, consider whether you want an evergreen or deciduous one. The former will provide year-round screening, but you might feel you only require screening in summer, when you spend more time outside in the garden. Deciduous hedging provides more impressive autumn colour than evergreen plants, as well as attractive spring blossom and colourful berries.

Mixed hedging using both evergreen and deciduous shrubs is often a good compromise if you want the best of both worlds. Choose plants such as elaeagnus, griselinia and holly for evergreen structure, and intersperse with deciduous hawthorn, field maple, guelder rose, blackthorn etc. for a range of berries, autumn interest and to attract wildlife.

Hedge growing hacks

- Choose young shrubs when planting a hedge as they establish quicker.

- Bareroot trees (not in a container) are inexpensive, but only available during winter.

- Water regularly in dry weather during establishment in the first year, making sure you provide enough water to seep down to the roots, where it is needed.

How to plant

For a really dense screen that can double as a windbreak or sound barrier, then it is best to plant in two parallel rows, one behind the other, in a zigzag formation. This will create a dense screen. Use two parallel strings to mark out the two rows for the hedge at 30cm (12in) apart. Plant the first row at 60cm (24in) apart between each shrub. Then plant the row behind, also at 60cm (24in) apart but starting 30cm (12in) in from where the first row was started. Single rows are fine if a only a simple screen is required, for example where a garden boundary or seating area needs defining.

Right: A high, dense hedge creates an attractive, useful and natural backdrop to a seated area.

Weave a willow or hazel screen

Weaved fences can transform a small plot into a gorgeous, rustic-looking cottage garden. There are various plants you can use for weaving your own fence, but the stems need to be supple enough to bend and also need to be readily available. Hazel and brightly coloured willow are two of the most popular plants to use.

A colourful option

There are numerous brightly coloured willow stems that can be used, including black, yellow, purple, bronze and red, which add vibrancy to a garden design.

You will need:

Secateurs to cut the young stems

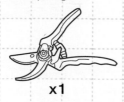

x1

Pruning saw

x1

Young stems of hazel or willow ('withies') about 1m (3ft) long

String

Thicker stems of hazel or willow for stakes, about 4cm (1½in) diameter

Lump hammer

x1

❶ Use secateurs to clean up the stems by removing any side shoots, as this will make weaving the fence much easier. You can save the side shoots for propagation/cuttings.

❷ Choose some thicker stems to create the upright support stakes. Cut them with a pruning saw to the height of the intended fence, allowing a third of the length for driving into the ground.

❸ Mark out with strings where you intend the fence to run. Then use a lump hammer to drive the upright stakes into the ground 30cm (12in) apart.

❹ Start in one corner at the bottom and weave the withies in between the posts. Trim off the ends so that the end of each withy is about 2cm (¾in) past the furthest stake it will reach.

❺ Continue weaving your way around the upright posts, alternating which side of the stakes you start from, as this will make a strong structure.

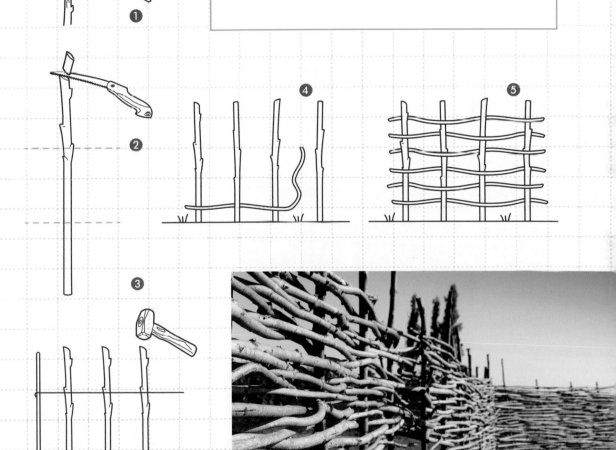

When to weave

Late winter or early spring is often a good time for weaving a fence, because it is when lots of gardeners are coppicing (pruning back to near ground level) their winter colour-stemmed plants. If you pop around to a local public garden you may well be offered a bundle to take home. Also, the branches don't have leaves at this time of year, which makes weaving easier.

Grow shade-loving plants in containers

Urban garden settings or those in tight, shady spaces present the ideal opportunity to introduce shade-loving plants grown in containers. They can be placed on steps up to a flat, in porches or in a small shady corner on the patio or balcony, brightening up even the darkest of areas.

One key benefit of shade-loving plants is that they are less likely to dry out, as they receive, and need, less sunlight. This means less watering and maintenance. And when growing such plants in containers, you can take them with you when you move. Containers can be fitted into the tiniest of spaces, and might be the only option if a garden has been concreted over and/or there is no access to the soil.

Choose your style

Finding containers that look good is just as important as the plants in this scenario, and helps to set the tone and design for your outside space. Terracotta pots are traditional and give a Mediterranean feel to an area. Wooden containers look rustic and natural, whereas upcycled options such as old bins, buckets or even wheelbarrows filled with plants can give a space individuality. Metal containers can look stylish and urban, and as they're in the shade, they won't overheat.

Choose where

If there is space for more than one container, position them in groups of odd numbers, such as three and five, ideally in a range of sizes, with the smallest in front and the largest at the back. A cluster like this will look more appealing than a straight line, placed in a corner or against a structure such as a wall, so they don't look randomly plonked.

Shade-lovers

For foliage – many shade-loving plants suitable for containers have impressive foliage, such as hostas, ferns, ivy, heucheras and wood spurge (*Euphorbia*); or for something larger, try a Japanese acer or a tree fern.

For winter interest – hellebores love light shade and make an attractive display in late winter. Other traditional woodland winter plants such as snowdrops, cyclamen and winter aconites look great in pots.

For summer flowers – try foxgloves, astrantias, cranesbill geraniums and aquilegias.

For texture – the upright foliage, flower plumes and seed heads of ornamental grasses can look stylish in pots. Either plant them individually or mix them up by contrasting them with some of the other plants mentioned. Suitable shade-loving grasses include snow rush *(Luzulu niviea)*, Japanese forest grass *(Hakonechloa macra)*, Carex *(sedge)* and big blue lilyturf *(Liriope muscari)*.

For something contemporary – try the dark grass black mondo *(Ophiopogon planiscapus* 'Nigrescens'), which is happy in partial shade. Place white pebbles on the surface of the compost to create a striking contrast.

From top left: Heucheras' varied foliage creates a tapestry of textures and hues.

Hellebores thrive in light-shady areas of the garden, and produce rose-shaped flowers in late winter and early spring.

Foxgloves are biennials and can be used to create a cottage-garden look in the smallest of shady gardens.

Black mondo grass is impressive in contemporary settings and small urban gardens.

Grow fruit trees in containers

Almost all fruit trees can be grown in containers and make beautiful features, providing both blossom and fruit. By growing them in containers they hardly take up any space, which is ideal if you have a small garden.

Fruit trees can be grown in almost any type of container. Consider reusing materials such as old buckets, dustbins, barrels or water butts, although you may need to create some drainage holes in the bottom with a drill. You can also make a container easily from recycled wood or pallets.

Of course, you can also plant a tree in a pot bought from the local garden centre; just ensure it is frost resistant. Ideally, the container should be about three times the size of the root ball of the tree you have bought. As the tree grows, you can repot it into something larger every few years.

Types of fruit trees to try

Of all the fruit trees, apple is probably the easiest to grow in a pot, as it is hardy and readily available to buy. There are hundreds of varieties to chose from, but it's nice to choose an unusual, local variety that you wouldn't find in a supermarket. Pears also make attractive specimens in a pot and are easy to grow.

Stone-fruit trees such as cherries, plums, damsons, peaches and apricots have attractive blossom as well as delicious fruit. As they flower early in spring, protect their blossom with a fleece if cold temperatures are predicted.

Three to try

1 A fig such as the variety 'Brown Turkey' adds a touch of the Mediterranean to any patio or balcony.

2 Cherry trees have abundant spring blossom and both sweet and sour cherries to choose from.

3 A lemon tree can be moved to shelter in a porch or conservatory when the weather turns cold.

Tree in a bin

Apple trees can even
be grown in old
dustbins (see overleaf),
just be careful the roots
do not overheat
and are regularly
watered if in a hot,
sunny location.

Plant an apple tree in a dustbin

An old dustbin is ideal for growing an apple tree in as they are cheap to buy, or free if you reuse an old one. Not only will it make a great feature in a small garden, but also, by having the tree's roots restricted in a pot, it ensures the fruit is at a nice height to pick, as well as being easy to maintain and prune each year.

You will need:

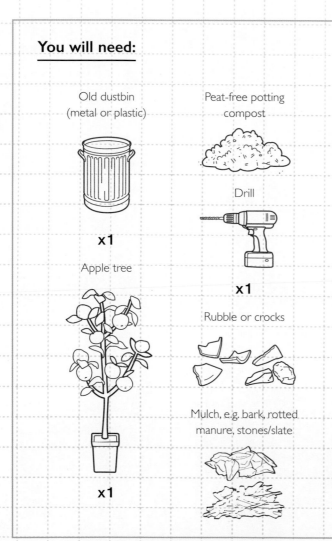

Old dustbin
(metal or plastic)

x1

Apple tree

x1

Peat-free potting
compost

Drill

x1

Rubble or crocks

Mulch, e.g. bark, rotted
manure, stones/slate

1 When growing an apple tree in a container, it is important to ensure there is adequate drainage to prevent the roots rotting. Use a drill to create 1cm (⅜in) holes about every 10cm (4in) in the bottom of the bin.

2 Place rubble or crocks over the drainage holes to prevent them clogging up with compost, then fill the dustbin half full using a good-quality potting compost.

3 Remove the tree from its original pot, tease out any roots if they are tangled and compacted, and place the tree in the new container, on top of the existing compost.

4 Ensure the top of the tree's root ball is just below the top of the container. Once satisfied, backfill around it with more compost until it is level with the top of the root ball, firming it in.

5 Mulch (cover) the surface of the compost with a 2cm (1in) layer of bark, rotted manure or stones/slate. These will help to retain moisture and suppress any weeds.

Basic container-grown fruit tree care

- Prune apple and pear trees in containers once a year using secateurs.
  Do this when the tree is dormant, for example in winter, when the leaves have
  fallen from the tree. Plum, apricot, peach, cherry and other stone-fruit trees
  should be pruned once a year in spring or summer, once the tree comes into
  leaf. This is to avoid disease.

- When pruning trees, remove about a sixth of the branches, including
  crossing branches and any diseased and dying wood.

- During the summer, fruit trees may need watering every few days.

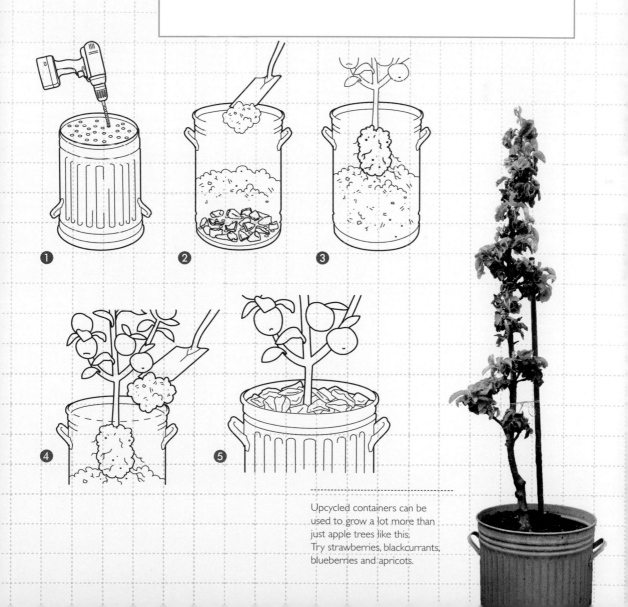

Upcycled containers can be
used to grow a lot more than
just apple trees like this.
Try strawberries, blackcurrants,
blueberries and apricots.

28

Grow fruit in hanging baskets

Hanging baskets can be used to grow colourful, tasty fruit and are a great way to brighten up a small garden. If short of space, they can be hung outside the backdoor or kitchen window. Not only are they at an easy height to pick, but also, hanging baskets keep tempting berries away from slugs and snails.

The best types of fruit to grow in a hanging basket are ones with trailing habits such as strawberries, cranberries and tomatoes. Growing them in a basket saves using up precious ground space.

Strawberries and tomatoes

The red, juicy, succulent fruits of strawberries and tomatoes tumbling down from a hanging basket look beautiful, as well as tasting delicious. Strawberries will benefit from a few plants in the basket to make it full. Plant one in the centre and four more equally spaced around the sides. The best type of compost to use is peat-free potting compost.

For something a bit different, try growing pink-flowering strawberries such as 'Pink Panda', as you will be rewarded with gorgeous-coloured blossom in spring, as well as delicious fruit later in the season.

Tomatoes are also suitable for hanging baskets, and only one plant is needed for an impressive display and a bumper crop; choose a variety with a trailing habit, such as 'Tumbling Tom Red'.

Cranberries and lingonberries

Cranberries provide tart-tasting red berries in late autumn and into winter, perfect timing for impressing family and friends by using them to make homegrown, homemade cranberry sauce for Christmas dinner. They require acidic soil so should be planted in peat-free ericaceous compost, which is available at garden centres. Alternatively, make your own by mixing homemade compost with rotted pine needles in a 50:50 ratio. Cranberries also prefer slightly moist conditions, so line the base of the hanging basket with a recycled plastic bag punctured with a few holes to prevent the water stagnating. Remember to water regularly and keep moisture levels topped up.

If you want to grow something more unusual and difficult to find in supermarkets, try the tart-tasting, bright red lingonberry. Sometimes called cowberry; it has similar growing requirements to cranberries.

Left to right: Position hanging baskets or troughs in full or partial sun but sheltered from the wind, so that they don't sway around too much.

Strawberries will need watering almost daily in early summer to ensure juicy, ripe fruit.

Feed tomato plants once a week during the summer with a tomato feed and water every few days to encourage good yields of plump, well-ripened fruit.

Cranberries should be watered with rainwater instead of tap water, so as not to alter their acidic compost.

Harvest strawberries from hanging lengths of guttering

Growing strawberries in lengths of plastic guttering is a practical space-saving solution to growing delicious fruit at home. It is a useful technique if you don't have suitable soil or much space, as the gutters can be suspended from fences or porches. Carefully positioning the guttering at a comfortable height will ensure it's easy to pick the fruit and to keep it weed free. Using deeper guttering will help to ensure that roots have enough space and stay moist in the summer.

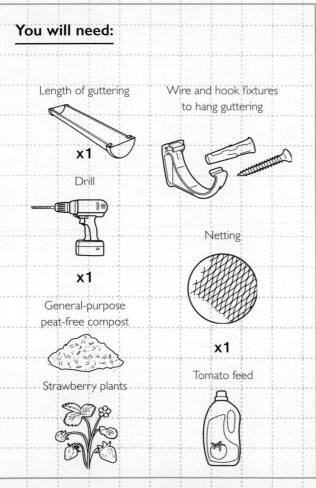

You will need:

Length of guttering
x1

Drill
x1

General-purpose peat-free compost

Strawberry plants

Wire and hook fixtures to hang guttering

Netting
x1

Tomato feed

❶ Take a length of standard household plastic guttering and drill small drainage holes along the base of it at a spacing of about every 10cm (4in).

❷ Fill the length of guttering with compost to just below the top, and place a strawberry plant in the compost every 20cm (8in).

❸ Position the guttering on a table or suspended between two chairs, logs or whatever you can find. It may be necessary to stabilize the guttering with two bricks on each side of the guttering at both ends, to prevent it rolling over. Alternatively, suspend the guttering from a pergola or attach it to a fence with hooks and wires. Drainage can be improved by positioning the guttering on a very slight incline so one end is slightly lower than the other.

❹ Place a net over the fruits as they start to ripen, to prevent birds from taking the fruit, and water the plants daily during summer. Feed with a tomato feed once a week.

Cultivate the stragglers

Cut the straggly runners that hang down from strawberry plants with secateurs and plant in compost. The following year, plant into guttering to provide another year of delicious fruit.

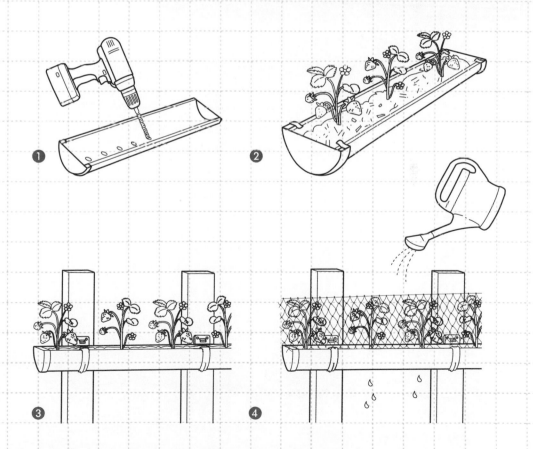

29

Grow a cut-and-come-again buffet on your windowsill

Growing cut-and-come-again salad leaves is easy, and the tiny leaves take up very little space, making them perfect for a window box. The biggest advantage to growing them is that they can provide a continual harvest of young, tender leaves over a longer period, instead of a single glut at the end of the season.

Cut-and-come-again leaves, like the names suggests, are cut with scissors or a knife when required, leaving the remaining plant to sprout, or 'come again', a few days later, ready for harvesting again. It is important when cutting to leave about 2cm (1in) of the plant so that it can shoot again.

What to sow

There are many different types of leaves to choose from. You can buy pre-mixed packets of seeds, or you can choose your individual favourites and mix them up or sow individual rows of the chosen salad leaf. Suitable varieties for sowing include:

Beetroot, chicory, coriander, chard, endive, cress, leaf celery, lettuce, mizuna, mustard, pak choi, parsley, purslane, radicchio, rocket, sorrel, spinach.

How to sow

In spring, fill a window box full of peat-free compost and sow seeds in shallow drills about 5mm (⅛in) deep, 1cm (½in) apart between each

seed and 12cm (5in) between each row. A typical window box can probably accommodate two rows. Lightly cover the seeds with compost. It can help to water the drill first, to help the seeds stick. Water the seeds after sowing, too, using a watering can with a 'rose' fitted over the nozzle.

Place the window box on a sunny window ledge and water regularly.

Note: If sowing some of the oriental types, such as pak choi, komatsuna and mizuna, it is best to wait until summer to start sowing, as earlier sowings can cause the plants to flower too early and go to seed, at the expense of producing lots of leaves.

Keep it going

Continue sowing small batches of seeds every two weeks up until late summer. Each sowing can be cut three or four times, before the leaves become bitter and tough. Just pull up the plants and add them to the compost heap. Top up the window box with more compost and sow again.

Reap then eat

Eat the leaves within a few hours of picking for optimum freshness. To extend their shelf life, splash the leaves lightly with water and place in a bag in the fridge.

Cutting delicious, fresh salad leaves whenever needed ensures there is a steady but manageable amount for using in the kitchen.

Above left and right: An abundance of tasty, attractive leaves can be grown in the smallest of containers – pack them in and pick them regularly to keep up production.

Sow chia seeds

Chia seeds are considered by many to be one of the healthy 'super foods', and can be bought from food shops and supermarkets to add to smoothies or even glasses of water as a nutritious drink. They're also used in baking, and give a light crunch to salads, casseroles and desserts. A member of the mint family *Salvia hispanica*, they are very easy to grow either indoors or outside.

You will need:

Rake (if growing outside)

x1

Seed tray (if growing inside)

x1

General-purpose peat-free compost

Watering can

x1

Chia seeds

x1 packet

Paper bags or jars for storing seed

Growing outside

Wait until the risk of frost has passed then choose a location in the sun with well-drained soil. Rake over the soil to make it friable, or crumbly. Add compost at a rate of about half a wheelbarrow load per square metre. Dig it into the first few centimetres below the surface of the soil, then lightly scatter the seed and rake it in.

Leave the plant to produce attractive blue flowers for the wildlife to enjoy and collect the seed a few weeks later. Use some of the seed for cooking and save the rest for sowing fresh plants again next year.

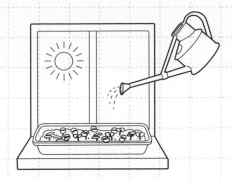

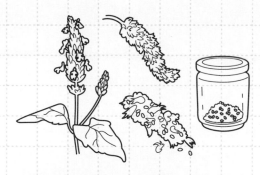

Growing inside

Chia seed can also be sown indoors all year round on a sunny windowsill. Fill a seed tray with seed compost, add water to make it slightly damp and lightly sprinkle the seed over the area.

After about two weeks, the seeds should start to germinate. Young micro leaves can be picked and used in salads as an alternative to young spinach leaves. Alternatively, dry or fresh leaves make a healthy tea.

To harvest the seeds

Wait until the flowers have finished and start to turn brown. Place them in a paper bag and leave them to dry for a few days. Then shake the seeds from the flower heads and store them in a jar until ready to use.

Where to buy

Buy a packet of chia seeds from a supermarket or health food shop. Alternatively, seeds can be purchased online or from some garden centres.

30

Create an indoor jungle

Surrounding yourself with lush, tropical-looking plants has become fashionable when it comes to designing an interior. Being in a room with lots of glossy foliage and brightly coloured flowers is a great way of embracing your adventurous spirit, making you feel as if you have been transported to somewhere far away and tropical.

Enjoying a taste of the exotic from the comfort of your own home has never been so easy. Whereas outside in the garden you are confined to jungle-like plants that are tough enough to cope with cooler weather; indoors there are no such limits. Practically anything that grows in a jungle can be grown in a heated house or apartment.

Contrast foliage

Creating the jungle look is all about foliage. The trick is to feel completely immersed in the plants. Look to fill wall space from floor to ceiling with big, bold exuberant planting.

Use upright stems of bamboo-like plants to contrast with large-leaved ones, such as a cheese plant, as if they are jostling for position in the wild. Run climbing plants such as bougainvillea or hoya up through the foliage, encouraging them to scramble among each other.

Use plants with a trailing habit, such as spider plants or string of hearts, to hang downward from high shelving. Also consider draping Spanish moss among the foliage. It doesn't require soil or compost, but it will need spraying with water every few days to keep it alive.

Once you have the foliage plants in place, you can try some flowering plants in front of them, such as the impressive bird of paradise, orchids or some bromeliads.

Water wisely

Whereas the main cause of death for plants growing outside is lack of water, very often it is the opposite with interior plants: overwatering kills most of them. Always read the plant label to check requirements. Indoor plants in containers should be placed in saucers to catch any excess water, and these can occasionally be topped up to water the plants from the bottom. Except for a few foliage plants such as Spanish moss, which like to have their foliage sprayed, many jungle plants should be watered from underneath, via their saucer.

INDOOR JUNGLE PLANT SELECTOR

There are hundreds of bold and luxuriant indoor plants that can give you that jungle theme in your favourite room. Pick ones with large, dramatic foliage or striking bright flowers to feel as if you are somewhere exotic. Aim to fill your indoor space with as many as possible for the full jungle experience.

AMAZON ELEPHANT EAR
Alocasia × amazonica

This luxuriant foliage plant will transform your room into a jungle-like atmosphere. It has large glossy leaves with spectacular silver veins running across them, and grows up to 1m (3ft). It prefers dappled shade or filtered light, needs feeding regularly and enjoys a rich but well-drained compost.

SWISS CHEESE PLANT
Monstera deliciosa

Producing giant glossy leaves, this plant is also a great air purifier. Leaves start off as heart-shaped but develop the trademark holes as they grow older. In the wild these plants are climbers, reaching up to 20m (66ft) high, but even at home they can reach over 2m (7ft).

BIRD OF PARADISE
Strelitzia reginae

Possibly one of the most recognizable flowering house plants, due to its distinctive blue and orange flower shaped like a tropical bird. It has attractive lush foliage, is fast growing and will eventually reach up to 1.8m (6ft). It likes sunny conditions and will need regular watering.

KENTIA PALM
Howea forsteriana

No jungle-themed room is complete without some sort of palm. The Kentia palm has dark green, glossy, arching fronds. It grows slowly up to 1.2m (4ft) and is low maintenance, meaning it will tolerate some neglect. It prefers bright conditions but out of direct sunlight. Water every couple of weeks.

BENJAMIN TREE
Ficus benjamina

An attractive, glossy foliage plant with either green or variegated foliage. Growing up to 1.8m (6ft), it can be fussy if near a draft, causing it to drop its leaves. It prefers indirect light and moist, humid conditions, so keep it away from radiators, which can be too drying.

SWEETHEART PLANT
Philodendron scandens 'Micans'

This is an easy-to-grow house plant with heart-shaped foliage that is pinkish when young. A natural climber that grows up to 1.2m (4ft), it can add to the jungle theme if left to scramble among other plants, but can also be placed on a high shelf and be allowed to trail downward.

31

Build a living wall or room divider

Dividing a room into different sections can create cosy nooks and provide privacy. The areas that dividers create can be used as work/creative spaces hidden away from day-to-day living. Best of all, though, dividers provide vertical opportunities to pack more gorgeous plants into the house.

Creating a room divider or screen can be inexpensive, and it is something you can do easily yourself. Living walls, dividers and screens can be temporary and even moveable if you place them on wheels.

Selective privacy

The nice thing about a screen as opposed to a permanent wall structure is that you can provide some privacy without detaching yourself entirely from other members of the household. Perfect if you have children you want to keep an eye on, or if you don't want to be excluded from flatmates' conversations on the other side of the screen.

Above: Bring the outdoors inside with a vertical plant divider, which provides a welcome contrast and a splash of nature in an otherwise urban setting.

Moss art

For something really low maintenance, choose preserved moss panels to hang on the dividers. These are framed arrangements of moss (and ferns and other greenery) that hang like pictures. Although technically not 'living' plants, they add a touch of serenity and greenery to an interior. They don't require watering, and don't even seem to collect dust.

Dividers are also useful for hiding clutter and paperwork. Best of all, they offer two blank screens, the back and the front, on which to grow plants.

Buy or make

There are two options when considering a divider. Firstly, dividers can be bought, then plants and structures fixed to them. This is useful if you don't rate your DIY skills or have much time.

Secondly, you can make them yourself using three sheets of marine ply. Add hinges to fix them together, so that they can be concertinaed on the floor, to keep them upright and sturdy. Paint with a waterproof paint to fit in with the surrounding decor.

Add plants

Fix brackets or shelves for plants to be placed in or on. Hanging bracket supports can also be used. Alternatively, light plant pockets can be bought made from canvas or Oxford fabric, which can easily be fixed to cover the dividers.

Choose trailing and climbing indoor plants to cover the screen, but also use orchids and bromeliads to add colourful flowers and provide an exotic touch.

How to water

Always check light and water requirements for plants. Ideally, choose plants that require minimal watering. If you are going to be watering regularly it may be worth putting temporary coconut-fibre matting down on the floor to soak up any excess moisture that drains out of the containers. Alternatively, place a tray on the floor beneath the divider when watering, to catch the drips.

32

Create an indoor hanging garden from cuttings

Many plants are easy to propagate from cuttings. It is worth giving it a go. Simply snip them off a plant and pop them in jars of water. Then suspend the jars of water around the house from hooks and shelves to create your very own 'hanging garden'.

Hanging or suspending a jar from the ceiling will not only save you space on your shelves and windowsills, but also, having the plants in your eyeline will make an impressive feature in an otherwise empty space. Taking a cutting from a plant couldn't be easier – most plants 'want' to be propagated. It is their method of reproducing, so by taking a cutting you are helping your plants out, as well as providing new plants for yourself.

How to take a cutting

Remove a healthy young shoot from an existing plant (a 'mother plant') using secateurs or scissors. The length of the stem should be 5–20cm (2–8in), depending on the size of the original plant. Cut the bottom of the stem just below a bud, to stimulate the plant to send out roots. Remove two thirds of the lowest leaves and insert this section into your jar of water. The plant will take a few weeks to send out new roots, providing you with a new plant.

Five to try

❶ Philodendron

❷ Tradescantia

❸ Coleus

❹ Pothos

❺ Syngonium

What to use

Most plastic or glass bottles and jars will do, as long as there is enough water for about two thirds of the stem to sit in. It's a great way to recycle old wine or beer bottles, or surplus coffee mugs. However, it is easier to see whether the plants are producing roots if the vessel is transparent.

Tie garden twine securely around your jars and bottles, ensuring they are supported at their base, and hang them from hooks around the house. Change the water every few days, as the older liquid can start to go brown and a bit stagnant. Once they have established roots, transplant the newly created plant carefully into a pot with drainage holes and containing potting compost. Alternatively, make a kokedama ball to continue your hanging garden theme (see page 114).

Left: Golden pothos (*Epipremnum aureum*) has attractive foliage and is easy to take cuttings from.

Make kokedama balls

This simple technique of suspending plants originates from Japan. The literal translation of *kokedama* is 'moss balls'. They are a wonderful way of incorporating house plants into your living space without taking up too much space. They make impressive green, living architectural features when suspended from the ceiling, curtain rails or hooks attached to the wall.

❶ Mix together equal parts of the bonsai and multi-purpose compost in a mixing bowl. Gradually add water to make the compost mix sticky, then mould it into a ball about half the size of a football.

You will need:

Bonsai compost

Multi-purpose peat-free compost

Sheet of moss

x1

Mixing bowl

x1

Garden twine & scissors

Shade-loving perennial such as an asparagus fern or phalaenopsis orchid

x1

Finding moss

Sheets of moss to hold the kokedama ball together can be purchased online and from specialist shops. Alternatively, if you know somebody with a mossy lawn, a spring-tined rake can be used to scrape up moss and bind the ball together.

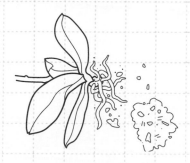

2 Remove your chosen shade-loving perennial from its pot and gently tease out some of the roots, removing some of the surplus compost.

3 Break open the ball of compost you have made into two halves or make a hole in the ball. Place the roots of the plant carefully in the gap and mould back into the shape of a ball.

4 Lay the sheet of moss out on a table and place the ball on it. Wrap the ball up in the moss, ensuring that the foliage of the plant is protruding out of the top or sides of it.

5 Secure the moss by wrapping garden twine around it. Hang it somewhere inside, out of direct daylight as moss balls dry out quickly. Spray the ball with a water mister regularly to keep the moss and plant moist.

Colour your garden with annual flowers

There is nothing so quick and easy in the gardening world as sowing annual flower seeds. You can fill an entire garden with these beautiful blooms for the cost of a few seed packets. Annual flowers provide a spectacular burst of colour for one season. Not only will you enjoy them, but the wildlife will benefit from the flowers and seed heads, too.

Annuals, as the name suggests, live for no longer than one year. They are usually sown in spring (sometimes autumn), and will grow, flower, set seed and die all in the space of a few months. Most vegetables are annuals, but there are also some beautiful flowers worth trying.

The main advantage of sowing hardy annual seeds is that the rewards are fast. As annuals live, flower, set seed and die within one season, they need to grow quickly because their time is limited. Annual flowers tend to have the brightest flowers, to attract pollinators in their small window of opportunity.

Clockwise from top: Plant a good mix of annual flowers for a kaleidoscope of colour.

Sow sweet peas into toilet-paper rolls and plant out directly into soil, where the cardboard will break down.

Collect the seeds from the seed heads of poppies to sow next year.

Choose your flowers

Hardy annual seeds with beautiful flowers include:

Marigold (*Calendula officinalis*), cosmos (*Cosmos bipinnatus*), poached egg flower (*Limnanthes douglasii*), sweet pea (*Lathyrus odoratus*), sunflower (*Helianthus annuus*), nasturtiums (*Tropaeolum majus*).

There are many others to choose from if you browse seed catalogues or online. They often have the abbreviation 'HA', for Hardy Annuals. Another advantage of buying seeds online or from catalogues is that seed packages are very light, so postage costs are minimal.

How to sow

Sow from early to mid-spring. Dig over the soil lightly to remove any weeds. Rake it into a fine tilth, then scatter seed lightly over the area according to the sowing rates on the seed packet – often it is a small handful per square metre. Lightly rake the seeds in so they are just under the surface of the soil. Keep them watered during dry spells in the next few weeks, and a short while later you will be enjoying your very own beautiful flower patch.

Collect the seed

An additional benefit is that seeds can be collected at the end of the year for using the following year. Store in envelopes in a dry, dark place, such as a drawer, then sow the next year. Some seeds might not be exactly the same from year to year, but it is exciting to see what variations grow. Who knows, you might even create a new famous variety from your seedlings!

Sweet-pea pots

As sweet peas are climbers, sow seeds individually in pots then grow them up a wigwam structure (see page 76).

Sow a sunflower patch

Sunflowers come in a range of bright colours, including bronze, yellow, orange, red and brown. They are one of the easiest and most reliable hardy annual seeds to grow. Some of the taller varieties can reach the heady heights of 4.5m (15ft). For a bit of fun, compete with other members of your household to see who can grow the tallest.

You will need:

9cm (3½in) pots

Plant labels

General-purpose peat-free compost

Bamboo canes for staking

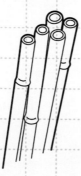

Sunflower seeds (a tall variety, i.e. 'Mammoth', or a multi-headed, branching type, i.e. 'Velvet Queen')

1. In spring, fill 9cm (3½in) pots with a general-purpose compost. Sow one seed into each pot at a depth of 1.5cm (¾in) and label the plants clearly.

2. Place the sown seeds on a sunny window ledge and keep them well watered. When they have reached a height of about 8cm (3in) they are ready to be planted outside.

3. Lightly dig over the soil where the sunflowers are to be planted. Add a bucket load of garden compost per square metre to the soil, to help sustain the plant as it grows. Space plants 45cm (18in) apart.

4. If it is a tall variety, the plants will need staking with sturdy cane supports to prevent them snapping in the wind. Keep the plants watered in dry weather as they grow.

5. Once a sunflower has finished flowering, don't cut down the flower head immediately, but instead leave it for the birds to enjoy. Save any remaining seed for next year.

Multi-headed sunflowers are
shorter than single-headed
types but produce loads of
flowers, which can be cut
for displays. New flowers will
keep emerging until the end
of summer.

Grow fruit from leftover pips or stones

Growing fruit trees from pips found in fruit from the supermarket is an easy and inexpensive way to start a garden. By removing them from the fruit and providing them with light, water and compost, they will grow into trees for free.

Each apple pip is potentially a tree in the making. The pips are the plant's seeds, and will grow into a full tree if nurtured with the right care and attention. If you are patient and up for the challenge, growing from pips and stones is an interesting way to provide yourself with a collection of practically free fruit trees.

There is a chance that once the seed has germinated and has grown into a mature plant it might not bear fruit. Even so, it will still make an attractive tree if planted outside, providing a habitat for wildlife. Plant in a container and you can take it with you if you ever move house.

What will the fruit be like?

If the seedling does produce fruit, it will probably not be the same as from the original tree it came from. For an apple tree to bear fruit, its blossom has to be pollinated by another apple tree, of a different variety, flowering nearby. So the apples on the new tree, in the same way that a child is not identical to one parent, will have some traits of both trees. However, the exciting thing is that this is how new varieties are created. Many apple trees that end with words such as 'pippin' and 'seedling'

(probably the most popular being Cox's Orange Pippin and Bramley Seedling) were just natural seedlings that happened to produce delicious fruit. So, try growing your own, and see if you create an exciting new and delicious variety.

Planting from pip

Remove the pips from the apple core and soak them for a couple of hours in water. Then place them into general-purpose peat-free compost in a small pot, and place on a sunny windowsill. After a few weeks, growing tips will appear, meaning they are ready to be planted directly into the soil outside or placed in a larger container.

Growing from stone

Pears and many stone fruits, such as cherries, plums, damsons and peaches, can be treated in exactly the same way.

If you want to try something more exotic, try growing a mango or avocado from the large seed husk found inside. They make beautiful house plants and might even fruit – if you have a warm conservatory and are willing to wait a few years!

Clockwise from top left: Germinate an avocado stone by soaking the base in water for a couple of weeks.
Try growing a glossy-leaved mango tree (*Mangifera indica*).
Apple pips can be sown into compost, or soaked for a few days first to sprout.
When apple seedlings reach 10–15cm (4–6in) high they can be planted outside.

Grow tomatoes from seed

Harvesting fresh tomatoes that you have grown yourself is
extremely rewarding, and they taste sensational when picked fresh
and warm from the vine. Instead of paying for packets of tomato
seeds, you can just take a standard bought tomato, slice it up, and
use the seeds inside to grow into an abundance of plants that can
be grown on a sunny patio.

You will need:

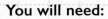

Tomato

x1

Sharp knife

x1

Chopping board

x1

Shallow container
or seed tray with drainage holes

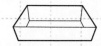

x1

General-purpose peat-free compost

Tomato pots or containers (9cm/3½in)
x several

Large pots (15 litres/3 gallons)
x several

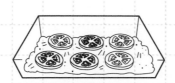

1 In early spring, take any tomato that you have bought from the supermarket or grocery store, and cut it into 3mm- (⅛in-) thick slices using a knife and chopping board.

2 Fill a shallow container or seed tray with compost and lay the slices of tomatoes on top. Space them out so that there is about 2.5cm (1in) between each one.

3 Cover the tomatoes with a 1.5cm (½in) layer of compost and water. Place the tomato seeds on a sunny window ledge and keep them watered as they germinate and start to grow.

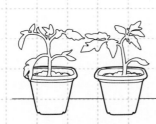

4 After a few weeks, seedlings will start to emerge from the soil. When they are about 5cm (2in) high, carefully separate them from one another and plant them individually in pots.

5 Once the risk of frost has passed, plant the seedlings outside in larger pots filled with compost, in a sunny, sheltered location. Alternatively, plant them in a conservatory or indoors by a window.

Fill kitchens and bathrooms with luscious foliage

Create a wow factor by filling your bathroom and kitchen with plants that will thrive in the shade and moisture these rooms typically provide. Introducing feature foliage and hanging plants will really bring these practical rooms to life.

Not only can the various textures and colours from the foliage of house plants improve the aesthetics of practical rooms such as the bathroom and kitchen, but they also help to purify the air as they absorb odours and carbon dioxide, and pump out clean oxygen.

Being surrounded by plants can also improve your mental wellbeing, which can help you relax in the bath or shower, or destress when cooking in the kitchen.

How to choose your plants

It's important to select your plants carefully so that they thrive in the often humid and shady conditions of these rooms. Thankfully, these are the same conditions as the lush, warm conditions of the tropics and the floor of the rainforests, meaning there are plenty of exciting plants to choose from.

As bathrooms and kitchens are often small spaces, you may want to consider the size of a plant, too. While large, dramatic architectural plants will create a focal point or feature, consider also using plants that can fill narrow shelves or fit on the side or corner of the bath. Trailing plants are useful as they can hang down from the sides of your cabinets, or even be draped over the shower rail, such as Spanish moss (*Tillandsia usneoides*, see page 127).

You could also consider exciting and very cool-looking carnivorous plants (see page 128), such as Venus fly trap, and pitcher plants in some of the rooms. These not only look great but will also help keep insects and bugs away from your food in the kitchen. Do bear in mind, though, that they will require some natural light.

Right: House plants do not need to have bright flowers. Some of the most popular and impactful ones have dramatic and architectural evergreen foliage.

A tropical vibe

Choose attractive containers that fit the theme/style of the room, putting large plants at the back and bringing smaller ones to the fore. Arrange plants near a window, if possible, where there is more natural light.

DAMP SHADE PLANT SELECTOR

It's time to debunk the myth that kitchens and bathrooms are too problematic for house plants due to their atmospheric conditions. There are, in fact, plenty of plants suitable for the steamy, humid and sometimes shady environments of these rooms. Do remember that a spare bathroom will be less damp and humid than a bathroom in regular use, and choose plants accordingly.

SPIDER PLANT
Chlorophytum comosum 'Variegatum'

One of the easiest and simplest plants to grow, the spider plant has a trailing habit and is therefore perfect for placing on a high shelf or on top of a cabinet, where its masses of variegated, strap-like foliage can cascade over the edge.

SWORD FERN
Nephrolepis exaltata

This plant originates from the swamps and rainforests of South America and the West Indies, so is perfect for the damp, humid conditions of the bathroom and kitchen. It has gorgeous, arching, green foliage that can spill down the outside of the bath or from shelves and kitchen units.

BAR ROOM PLANT
Aspidistra elatior

This plant is as tough as a pair of old gardening boots and will tolerate neglect and shady conditions. It has attractive dark foliage reaching up to about 60cm (24in) and is suitable for perching in the corner of the bath or on the floor, where it can conceal unattractive bathroom features such as toilet brushes and weighing scales.

PEACE LILY
Spathiphyllum wallisii

This plant loves warm, humid atmospheres and has striking white flowers and attractive green glossy leaves called spathes. The plant helps to purify the air and will tolerate shady conditions. It is called the peace lily because the flower was thought to resemble the peace flag. So, you can have lovely peaceful thoughts when relaxing in the bath.

SPANISH MOSS
Tillandsia usneoides

Add a touch of the tropical rainforest to your bathroom or kitchen with this plant's quirky and unusual trailing habit. A member of the bromeliad family, it has no root system but instead is an air plant, absorbing moisture and nutrients through its long, narrow, green-grey foliage. Try draping one over the shower rail for a dramatic foliage effect.

MOTHER-IN-LAW'S TONGUE
Sansevieria trifasciata

A striking foliage plant originating from Nigeria, with upright, strap-shaped leaves, is very forgiving to neglectful owners. It is one of the few succulents that will tolerate some dampness and humidity in the bathroom or kitchen – but don't overwater it or deprive it of light.

36

Cultivate a carnivorous display

Most of us are probably familiar with Venus fly traps, but there are numerous other types of carnivorous plants that can be grown as a fascinating collection. Furthermore, these quirky-looking plants make a great talking point with friends and family, providing plenty of intrigue.

By following a few simple rules, carnivorous plants can survive either indoors or outside. If you provide them with the correct soil conditions, light and shade, they will thrive.

You can make impressive, bold displays of carnivorous plants indoors, using their various contrasting foliage and growth habits to great effect. Place them on shelves or windowsills according to their sunlight requirements. Create displays on well-lit kitchen tabletops and living-room coffee tables using larger plants as a centrepiece and smaller ones to surround it.

Correct care

Carnivorous plants prefer acidic and slightly boggy soil. Try to buy a specific low-nutrient compost formulated specifically for this group of plants. If you can't get hold of this, then ericaceous compost will suffice. If you grow them in containers, then place a perforated waterproof membrane or even just a repurposed plastic bag in the bottom, to help retain the damp conditions they require.

Rainwater

Most carnivorous plants should be watered with rainwater; tap water can alter the soil's pH and dissolved chemicals can be toxic to the plant. If you don't have a water butt, collect rain in a bucket. Never fertilize or feed carnivorous plants, as they get all their nutrients from the insects they trap.

Plants to try

Easy plants to try include Venus flytrap (*Dionaea muscipula,* which have sensitive pads that close on a fly when it lands), sundew (*Drosera,* which have a sticky surface that insects stick to) and pitcher plants (*Sarracenia,* which have funnel-shaped flowers that trap insects). Keep them in a warm, bright location in summer, and move them to somewhere shadier and slightly cooler in winter, as they need a period of dormancy. Remember to keep them topped up with rainwater. In mild areas, some of these species can be grown outside all year round.

For something a bit more exotic, try a tropical pitcher plant (*Nepenthes*). Grow it in a hanging basket inside the house, away from direct sunlight. Bathrooms are ideal because of the humidity and darker conditions. Try hanging one toward the back of the shower rail – just don't allow the tap water to come into contact with the compost.

Left to right: Venus flytraps (*Dionaea muscipula*) need to be kept moist, and will flower in the summer.

Trumpet pitcher (*Sarracenia* × *catesbaei*) is a quirky and architectural plant suitable for a sheltered garden.

Pitcher plants (*Nepanthes*) are dramatic trailing plants with funnel-shaped flowers that trap insects – ideal for indoor hanging baskets.

Sundews (*Drosera*) are one of the largest groups of carnivorous plants, small in size with beautiful, glistening traps.

37

Tend a terrarium display

The beauty of these indoor container miniature gardens is that they require minimal effort to create, and hardly any maintenance once they have been planted. They make an eye-catching feature in any room.

Right: Stylish geometric terrariums such as this pyramidical one can be centrepieces in their own right.

Below: Repurpose spaghetti and pasta storage jars to show off a miniature indoor forest scene.

Terrariums need to be transparent to let in the maximum amount of light, which also lets you admire the plants from all angles. Glass or plastic are the two most popular materials. Specifically designed terrariums, such as glass dodecahedrons, can be purchased, although you can also recycle containers from home. Bell jars are often used, or large glass jars. Other suitable containers include old aquariums, vases or even demijohns, although it is important that there is a big enough gap at the top to access the plants and maintain the environment. You may wish to place attractive stones in the terrarium, in which case you will need a reasonably large access hole.

A sealed terrarium

Many terrariums are sealed with a lid or bung, as this creates a closed ecosystem. The plants require minimal watering as they become self-sufficient; transpiration from the foliage condenses on the surface of the glass and trickles back down, providing water. This suits tropical plants, which enjoy the moisture being released from their large, lush leaves. These humid, closed terrariums can overheat in direct sunlight.

Suitable plants include: nerve plants (*Fittonia albivenis*), bush maidenhair fern (*Adiantum aethiopicum*), golden pothos (*Epipremnum aureum*) and strawberry begonia (*Saxifrage stolonifera*). There are lots of others – always check the label to see if the plant enjoys low light and warm, moist, humid conditions.

Gaining access

You may need a long-handled spoon, tweezers, tongs, chopsticks or even wooden knitting needles to be able to reach in and maintain the plants.

Create a sealed terrarium – place a layer of gravel on the bottom of your chosen container for drainage, mixed in with a handful of activated charcoal, to keep the atmosphere sweet and to avoid stagnation and fungal problems. Then add a layer of a 50:50 mix of compost and soil to a depth of 2.5cm (1in). Plant sphagnum moss in clumps in the compost, then fill with plants. The moss helps to retain and collect moisture, maintaining the humidity levels.

Open terrariums

Open or unsealed units, where plants are exposed to the air, are better for dry, arid-loving plants such as cacti and succulents. These containers can be placed on windowsills where the plants can bask in the sun, but do check on them in case the foliage starts to scorch.

Suitable plants include: sempervirens, sedums, echeverias and aloes.

Create an open terrarium – fill the base of the terrarium with equal parts of small gravel, sand and compost to a depth of 2.5cm (1in). Place rocks or stones in next, then plant the succulents and cacti into the compost.

Moisture check

To check the moisture level, look to see where the condensation level is on the jar. If it is less than a third of the way up, your terrarium may need watering. If it is right up to the top and dripping heavily, then it may be necessary to wipe around the sides of the terrarium, to mop up excess moisture.

38

Start a cactus collection

One of the most appealing aspects of growing cacti indoors is that they are low maintenance; they hardly need any looking after at all. Most of them originate from deserts and therefore are quite happy without being watered or fed for weeks or even months – ideal if you are often away from home.

Cacti are a favourite for beginners and seasoned gardeners alike because they are so easy to look after. There are so many different ones, coming in all sorts of shapes and sizes, so they are popular with people who like to collect things. Careful, though, as this hobby can become addictive! Due to their simplicity, they are also good for children to grow, although beware the spikes.

To start a cactus garden, it is important to understand the environment they inhabit in the wild. They thrive in dry, arid desert conditions with hardly any rainfall. They are also resilient – in the wild they are often in exposed locations and barraged with hot wind and sandstorms.

From left to right: An assortment of cacti in a varied collection of pots will adorn any windowsill.

The weird and wonderful shapes of cacti will keep you fascinated, and hook you into collecting more!

Cacti are not just green and spikey. If you look after them properly, they may reward you with a spectacular flower or two, such as on this prickly pear cactus.

Where to grow

Cacti like it warm and hot, so keep them on a sunny windowsill. They are perfect for filling this space, as most other house plants won't be able to tolerate the heat. Their preferred temperature is between 18 to 28°C (64 to 82°F), and keep them out of cold drafts.

Caring for cacti

In summer your cacti will need watering every two to four weeks, but in winter reduce this to once every couple of months. It is important to avoid overwatering cacti, as it can cause them to rot. Water the compost at the base of the plant and try to avoid soaking the cactus itself, as this can cause damage.

Mind the spikes!

When handling spiky cacti, it is best to wear thick gloves. Alternatively, they can be handled by folding newspaper over a few times to make a chunky strip of paper and using this to move them about.
The good news is that you don't have to handle them very often; as they are slow growing, they only need repotting every few years into fresh cactus compost.

Create an indoor cactus garden

There are thousands of cacti to choose from when creating a cactus garden. Try to select a range of different shapes and sizes to create an interesting and contrasting tapestry of colours and textures. Once you have planted them, there is very little else to do, except sit back and enjoy your handiwork.

You will need:

Drill and 4mm (⅛in) drill bit (if container doesn't have drainage holes)

Shallow container (about 10cm/4in deep)

x1

x1

Crocks or broken polystyrene

Cactus compost

Selection of cacti in 9cm (3½in) pots

1 With a drill and 4mm (⅛in) drill bit, create 1cm- (½in)- diameter drainage holes in the base of the container if there aren't any. Cacti are easy to look after, but the one thing that will kill them is excess moisture.

2 Place broken crocks or broken up pieces of polystyrene over the drainage holes. This will prevent the compost washing through the holes and helps with drainage.

3 Place a 2cm (¾in) layer of horticultural grit into the bottom of the container. This will help to increase the drainage of the container.

4 Fill up the container with cactus compost to about 2cm (¾in) below the surface and firm it down. Remove the cacti from their pots and gently tease out their roots.

5 Plant the cacti in the compost at the same depth they were in their pots. Firm them in, then top-dress the surface with more horticultural grit to the top of the container.

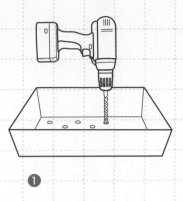

1

Shallow container

Cacti like a shallow container. It needs to be about the same depth as the pots the cacti are supplied in. Consider upcycling tins or plastic chocolate or biscuit tubs.

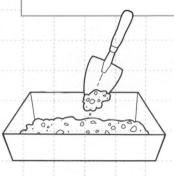

3

2

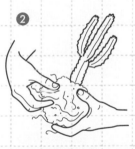

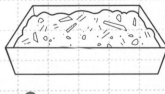

4

5

The container will likely need to sit on a tray so that the display can be watered from below.

39

Add spice to your kitchen

Spice up your kitchen by growing a ginger plant: not only does it make a beautiful house plant, but also, you can propagate it easily from a root bought at the supermarket. Keep the plant on a sunny kitchen window ledge and you will have this delicious fiery spice at your fingertips.

You will need:

Ginger root   Plant pot   Sharp knife

General-purpose
peat-free compost   Seed tray

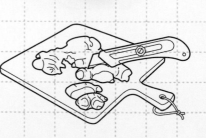

1 Once you have the ginger root at home, lay it on a chopping board and slice it into 5cm (2in) sections using a sharp knife. Ideally, each section will have at least one dormant growing shoot.

Who doesn't love the powerful, aromatic flavour of ginger? Whether you enjoy Asian-style cookery, baking gingerbread men or drinking ginger beer, this spice is a firm favourite with many of us. It is a root crop that originates from warm, tropical countries, but it will grow happily indoors on a sunny window ledge in cooler regions.

When buying ginger, look for plump sections of healthy root (technically it is a rhizome, which is a type of root) and discard any withered or dried-out ones. Inspect the root to see if there are any dormant growing shoots. These are knobbly lumps you can feel with your hands or see just below the surface that can potentially become a new plant.

Growing and harvesting

Plant your ginger root as detailed below. After a couple of weeks of growing, you will see the knobbly tips starting to expand and eventually emerge into a shoot. The root should be placed just below the surface, with the growing tip above it. Eventually it will develop into a beautiful, leafy house plant.

After a few months, the ginger root will be ready to harvest. Remove it from its pot and break off a section to cook with. Repot the remaining plant in fresh compost, or cut off more sections to produce new plants.

Tasty turmeric

Turmeric root can also be bought from the supermarket and grown in the same way as ginger.

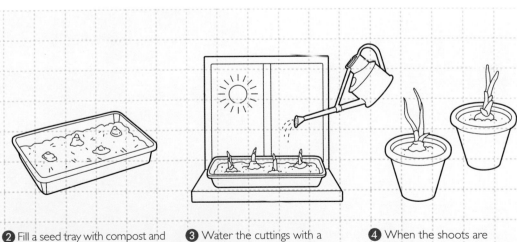

2 Fill a seed tray with compost and push the root cuttings into the soil so they are just below the surface, with the growing tips protruding.

3 Water the cuttings with a watering can with a rose over the nozzle, and keep the tray on a sunny window ledge.

4 When the shoots are about 8cm (3in) they can be removed carefully from the seed tray and potted up into individual pots.

Build a compost heap

To many gardeners, having a compost heap is the most important element in a garden. Not always the prettiest of features, if you have space for one, the many benefits to making compost will outweigh its looks. It can be used to improve the soil and fill raised beds, while recycling fruit and vegetable waste from the kitchen.

Making compost is easy, and a pile of compost doesn't have to be unsightly or take up much space. In fact, some composters are even designed to look colourful and attractive in their own right. There are also compact ones that take up hardly any room. Some composters can even be rotated by hand to save having to turn the compost by hand. They have secure lids and can be placed on a patio, ensuring that they are kept rodent-free.

Pallet composters

If there is space, a simple compost bin can be made by turning three pallets on their edges and screwing them together to form the two sides and a back, with an open front. Ideally, there should be three bins, one for leaving to rot down, one for current use on the garden, and the other for filling. Also, if there is more than one bin, you can have an empty one to turn the compost in to.

If you don't have space for a compost heap, turn to page 142 to find out how to build and use a wormery.

Get the mix right

The trick with compost is to get the right mix of green (nitrogen-based) material and brown (carbon-based) material. Aim for:

70% green: 30% brown.

Too much nitrogen will cause the compost to go slimy, whereas too much carbon will be too dry and it won't decompose properly.

Green material includes:

- Kitchen waste such as fruit and vegetables
- Grass clippings
- Soft herbaceous material from the garden.

Brown material includes:

- Shredded newspaper
- Cardboard
- Leaves
- Wood chipping.

Turning the compost

Compost material will break down gradually over a year, but if the heap is 'turned' every few weeks, the air that goes into the heap will speed up decomposition. Use a garden fork to move the material at the top of the heap and place it in a new heap at the bottom. Continue until the new heap has the oldest material at the top.

Clockwise from top left: The advantage of sealed or contained compost heaps is that they are rodent proof.

A healthy compost should have lots of worms, which will help decompose the material.

Open compost bins have easy access at the front, so it is possible to turn the heap regularly.

Make liquid fertilizer from nettles

Stinging nettles are full of nutritious goodness, and grow in abundance almost all year round. Despite their sting, these 'weeds' are a gardener's best friend, and can easily be made into a liquid feed. The nettles are packed full of nitrogen, a key ingredient needed to promote the healthy green growth of garden plants.

You will need:

A pair of gloves

x1

Black bin bag

x1

Scissors/secateurs

x1

Bucket

x1

Access to a patch of stinging nettles

Watering can

x1

Stirring stick

x1

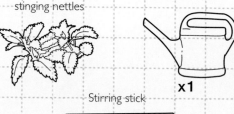

1 Use a pair of scissors or secateurs to chop and harvest a patch of nettles. You will need about half a bin bag full of foliage and stems.

2 Once back in the garden, the nettles can be chopped up into 15cm (6in) lengths with the scissors or secateurs. The smaller the pieces, the faster they will decompose in water.

3 Place the chopped-up nettles into a container such as a bucket or empty plastic bin and fill with water. The liquid will start to stink as it decomposes, so it is best to leave the bucket in a hidden corner somewhere out of the way.

4 After a few weeks the nettles should have rotted down to a brown sludge. Dilute the sludge in a watering can at 10 parts water to 1 part liquid nettle feed, and use a stick to stir it with.

5 Use this diluted liquid feed to water around the base of the plants during the growing season. This should be done once or twice a fortnight.

Switch to comfrey

Comfrey can also be used as a liquid feed if it is growing in the garden. It is made in the same way as nettle feed but is richer in potassium, so once a plant starts flowering or fruiting, switch to comfrey, and the potassium will help to promote colour and taste.

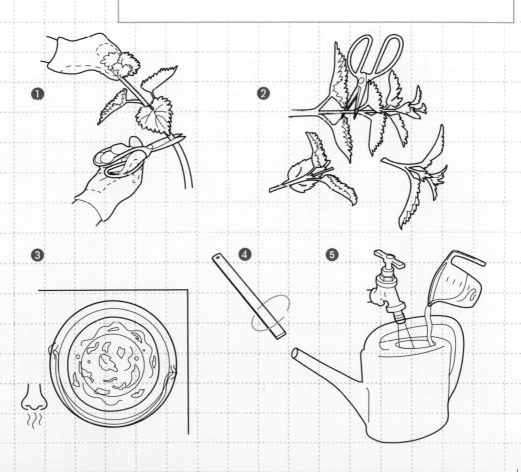

Fact

The technical name for cultivating a wormery for compost and liquid feed is vermicomposting.

41

Use a wormery

Worms are a gardener's best friend. Finding worms in the soil is a good sign, as it means the soil is healthy and fertile. They help to break down the soil and also aerate it as they move through it. By building a wormery, you can harness the beneficial qualities of worms and use their byproduct to create a liquid feed – 'black gold' – and a nutrient-rich compost.

When worms break down organic material such as kitchen fruit and veg peelings, they produce a lovely, dark compost as a byproduct, often referred to by gardeners as 'black gold'. This is perfect for using as a soil improver on the surface of flowerbeds. A liquid is also produced, which can be diluted in a watering can and used as a liquid feed.

Wormeries hardly take up any space, so they offer a practical solution for anybody who wants to compost their kitchen and garden waste yet doesn't have the room for a compost heap. Wormeries can be bought online or from garden centres, but it is easy to build your own.

Above: Feed a wormery regularly with small scraps of fruit and veg kitchen waste, and worms will transform it into compost for the garden.

Above right: Wormeries are a perfect, compact solution for providing compost on a balcony or porch.

The structure

A wormery is a small box-shaped structure with air holes, which is filled with green waste (see below) and garden compost. Once worms are added (carefully), they will break down the materials, leaving compost behind to be used in the garden. There is also usually a tap so the nutrient-rich liquid can be syphoned off.

For the feed filling the wormery, chop it up into small pieces to help the worms process it faster, and avoid adding too many watery or acidic products such as citrus fruits.

The worms

Different to those usually found in soil, the worms used in a wormery tend to live in decomposing manure, leaf litter or any organic decomposing matter. They are known under a few different names such as 'tiger', 'red wiggling' or 'brandling' worms, and are often reddish in colour with a yellow band around their 'neck'. Sometimes they are just referred to as 'composting worms'. If you look in a compost bin you will often find them there. These can be removed from an existing compost heap and placed in the wormery. Alternatively, packs of live composting worms can be purchased from garden suppliers. Once you have them, you shouldn't need to buy any more as they reproduce and multiply.

Ideal conditions

Worms are more active in the summer, so don't overfill your wormery during the colder seasons as the kitchen waste may just sit there and slowly rot. Wormeries don't like extreme temperatures, so as well as avoiding the cold, ensure there is some ventilation to keep it cool in summer. If outdoors, place a wormery in a sheltered position out of the wind. Wormeries should be warm and moist. Regularly check the material in the box and if it feels very dry, add a splash of water.

Dos and Don'ts

Products that can be added include:

- Coffee granules
- Fruit
- Cooked or raw vegetable
- Grass clippings
- Cardboard
- Leaves
- Weeds
- Herbaceous material from the garden

Don't add:

- Dairy products
- Meat
- Fish
- Eggs

42

Build a keyhole raised bed

Keyhole gardening is a clever technique that originated in southern Africa, combining a circular raised bed with a central compost heap. It is very easy to make, and if you use repurposed materials, it shouldn't cost anything. Keyhole beds make attractive, natural features in the garden, and can be made any size or shape, depending on the size of your outside space.

In the centre of a keyhole raised bed is a compost heap made from permeable material that leaches its nutrients out into the surrounding soil. It is called a 'keyhole' garden because the raised bed, if viewed from above, would resemble the shape of a keyhole, as it is circular with a narrow cleft. The cleft is a narrow walkway that allows access to the centre of the raised bed, where the compost heap sits.

Keyhole origins

The idea originates from southern Africa, in regions where the soil was unfertile and of poor quality. Raised beds were made from any free materials that could be found in the vicinity, usually rocks and stones, which would be stacked up in a circle to create the walls. The sides of the central compost heap would be made from reeds, grass or a thatched material that is permeable. By adding green waste regularly to the compost heap in the centre, the nutrients would leach out and into the surrounding soil. This provided the fertile conditions for vegetables, herbs and fruit trees to flourish.

Above: Leaving a walkway provides ease of access to the central compost, and allows you to tend to and harvest the rewards from your keyhole bed.

You will need:

Rocks, stones, wood or bricks

Organic decomposable material

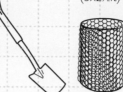

Spade

Chicken wire, 1x0.75m (3x2½ft) roll

Soil

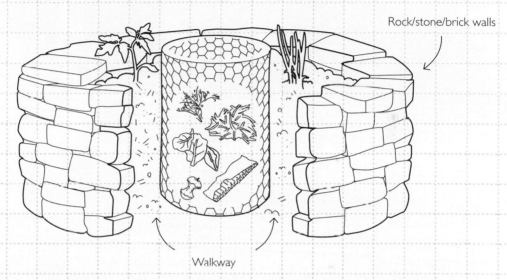

Rock/stone/brick walls

Walkway

1 Keyhole beds can be any size or shape you wish, but if it is too large the compost heap in the centre won't be able to supply the furthest areas of the raised bed with any nutrients. An ideal size, therefore, is about 2.5m (8ft) diameter.

2 Build up the walls of the circular raised bed to a height of about 50cm (20in). Suitable material can include rocks, stones, wood or bricks.

3 Leave a walkway to access the compost heap in the middle. From above, the structure should look like a huge round cheese with a slice taken out of it, or, indeed, a keyhole.

4 Use chicken wire to create a permeable 1m (40in) circular compost bin and fill it up with layers of any organic decomposable material, including grass clippings, fruit and vegetable waste and any other garden compost. Scrunched up newspaper and shredded cardboard can also be used.

5 Fill the raised bed with soil to just below the height of the walls. Ideally the soil toward the centre, next to the compost heap, should be about 10cm (4in) higher than out toward the edges so that maximum soil contact is made with the composting material in the middle.

6 Continue filling the compost in the centre as the season progresses. If the compost heap fills up, use wire cutters to make a small gap at the bottom of the chicken wire to take out the decomposed material and spread it over the beds as a mulch.

43

Create a no-dig vegetable garden

The popularity of no-dig gardening has increased enormously over the last few years. As the name suggests, it involves cultivating a garden without digging the soil. It is thought that this leads to a healthier environment for plants to grow in, as well as you needing to do less weeding.

The reason so many people love this method, apart from saving on hours of back-breaking work digging soil, is because it results in less weeding and a healthier, more fertile soil.

Exponents of the no-dig method argue that digging the soil damages its structure, which reduces its fertility, drainage abilities and results in less healthy micro-organisms. Also, digging into soil exposes and encourages previously dormant weed seeds to germinate. Finally, it increases the risk of accidentally chopping through the roots of perennial weeds, which can then create more of a problem, as it multiplies into more plants.

Instead of digging, a healthy soil can be created by building up layers of organic mulches on top of the surface. These layers gradually break down into a lovely, crumbly consistency, making it easy to pull out weeds, while creating a rich, fertile environment which in turn feeds the masses of soil organisms below the surface.

Where to start

Cut down any tall existing weeds to ground level using a strimmer, scythe or hand shears, avoiding digging into the soil. The cut material can be added to a compost heap as long as it hasn't gone to seed.

Cover over the soil with a sheet of cardboard. If it is a really vigorous or dominant patch of weeds, such as perennial nettles, then you may need two or three sheets. The idea is to smother out any light, as this will prevent the weeds returning, so make sure that the cardboard goes right up to the edges.

Next, add a 15–20cm (6–8in) layer of mulch on top. Use organic material such as garden compost, grass clippings, leaves, herbaceous material and fruit and vegetable kitchen waste.

This process does take patience, as this layer of organic mulch can take between six months and a year to decompose. Once it has turned into a lovely dark compost, it can be planted with vegetables.

Get growing

Use a trowel to dig out a hole just large enough to accommodate the seedling, avoiding disturbing the surrounding area. Alternatively, use a dibber to sow seeds, or a hoe to create a shallow drill.

If any weeds appear while the vegetables are growing, they should pull out easily from the loose crumbly soil without having to dig them out.

After the vegetables have finished cropping, pull up the plants and add them to the compost heap.

Add another layer of mulch to the surface of the bed, to a depth of 15–20cm (6–8in), and the no-dig cycle can start again.

Opposite: No-dig gardens should reduce the amount of weeds that appear in borders and beds.

Top: The key to successful no-dig gardening is to add compost regularly to beds to create a healthy soil.

Left: Plant directly into compost and the plants should thrive.

Build a mini Hügelkultur bed

Hügelkultur originates in northern Europe, and the literal meaning is hill or mound culture. It is a type of no-dig garden that involves cultivating crops such as fruit and vegetables on small mounds. These mounds are made of rotting logs and branches that are covered with organic material such as compost. As the wood slowly decomposes, it supplies the compost with nutrients that can last up to 20 years.

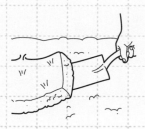

You will need:

Spade

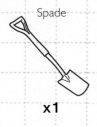

x1

Green, fresh material such as grass clippings, fruit and veg

Logs and branches — ideally already rotting, a variety of sizes, avoid anything diseased; or wood chippings

Rotted-down garden compost or general-purpose peat-free compost

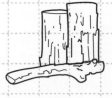

Vegetable plants — choose a range of plants to grow in shade and light

1 Identify where the mini hügelkultur bed is going to be made and start to prepare the area. Ideally it should receive at least some sun during the day. If there is grass, then use a spade to remove the turf.

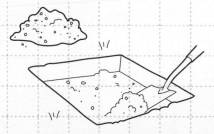

2 Dig out a trench about 30cm (12in) deep. It should be as wide and long as the longest branches and logs that you have found. A hügelkultur bed of about 2x1m (6½x3ft) is a good size for a small garden.

Sloping sides

Additional benefits of hügelkultur include increasing the growing surface area of a garden due to sloping sides. Some mounds are up to 2m (7ft) high.

Self-watering

Hügelkultur reduces the need to water as the rotting logs absorb moisture like a sponge and gradually release it.

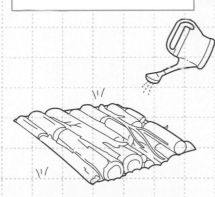

3 Place the logs and branches (or wood chippings) into the trench to fill it, so it is proud of the surface. Ideally, the logs will have already started rotting, as this will speed up the supply of nutrients to any plants. Water the logs to help the rotting process.

4 Start to build up the mound to an eventual height of about 75cm (2½ft) high. Begin with the grass turves that you removed earlier, placing them upside down on the logs. Next, add fresh material such as grass clippings and fruit and veg, then a thick layer (10–15cm/4–6in) of well-rotted garden compost.

The mound is ready for planting. If the sides are sloped slightly, they can be planted into as well as on the top (known as the plateau). Work out which sides are in the shade or sun and choose vegetables that prefer those conditions.

44

Plant a herb garden

Walking around a herb garden, rubbing the leaves to breathe in their volatile fragrances, is a relaxing and enjoyable experience. Easy to grow and suitable for containers by the kitchen door or in window boxes, having a wide range of herbs will increase your range of flavourings to use in the kitchen.

Herbs have been used for centuries for cooking, flavouring and medicinal purposes. Many are quite small, and therefore ideal for a smaller garden or courtyard. It is usually the leaves that are used, but depending on the plant, other parts such as the root, flower, seed, bark and sap can also be used.

Growing in ground or pots

Most herbs originate from dry, arid places such as the Mediterranean, and therefore prefer a warm, dry sunny location in the garden with well-drained soil. If you have heavy clay soil, then horticultural grit can be added to improve drainage.

Alternatively, herbs are easy to grow in containers. Choose a range of different-sized pots to make an attractive display. Frost-resistant terracotta containers look tantalizing with herbs spilling out from the tops and sides. Also consider using herb

planters, which have a number of planting side pockets, perfect for packing as many herbs as possible into a small area.

When planting herbs in containers, use a loam-based peat-free compost and add up to 25 per cent horticultural grit to improve drainage. Also place broken crocks or stones over the drainage holes in the bottom of the container. Place the container on bricks or feet to lift it off the ground and allow any surplus water to escape. Keep the herbs watered during a hot summer.

Hardy herbs

Some herbs are hardy and can stay outdoors all year round. These include bay, mint, oregano, sage, fennel, rosemary, chives and thyme. Some of the herbaceous herbs will die back when it gets cold and will only start regrowing when it warms up in spring. These herbs may need repotting into fresh compost every couple of years, if growing in containers.

Annuals and perennials

Other herbs are either annuals or tender perennials, which will perish at the end of the season. Plants include basil, coriander and French tarragon. These plants should be propagated each year in spring. Alternatively, they can be bought as small plug plants from garden centres and grown on a sunny windowsill in the kitchen. Growing them inside will extend their lifespan.

Contain your mint!

Mint is quite invasive if planted directly into the soil. It is far better to plant it in individual containers to keep it under control.

Above: Lots of different herbs with a range of tantalizing flavours can be squeezed together in beds.

Right: Plant herbs of different textures and heights in terracotta pots for a natural, earthy display.

HERB PLANT SELECTOR

Herbs are a wonderful addition to any garden. Not only do they have medicinal and culinary uses, but many are attractive, too. Parsley can be used to edge a path, while the distinctive purple or tricolour leaves on sage can be grown in a border to add texture and as a foil to surrounding flowers. There is beautiful upright or trailing rosemary with pretty blue flowers, and a bronze-leaved fennel that looks stately in the back of a border.

CHIVES
Allium schoenoprasum

A member of the onion family, chives are a perennial growing to about 40cm (16in) high and produce purple flower heads in summer. Plant them in full sun or partial shade in moist yet well drained soil.

MINT
Mentha

There are numerous species of this 30cm- (12in-) high perennial herb, and a range of nuances to the overall minty flavour that include apple, spearmint, ginger and even chocolate. It can be a bit boisterous in the borders, and is easier to manage if grown in a pot.

THYME
Thymus vulgaris

A low-growing perennial evergreen herb. They are fairly short lived and may need replacing every two or three years. Their tiny leaves are very aromatic and there are attractive silver and gold forms. A lemon-scented form is also available.

Flat-leaf parsley

PARSLEY
Petroselinum crispum

A popular annual herb, reaching about 30cm (12in) high and used in savoury dishes. There are usually two different types, curly or flat-leaved, both tasting similar but the latter usually having a stronger flavour. Sow each spring for a regular supply each year.

Curly parsley

FENNEL
Foeniculum vulgare

With beautiful, ornamental and feathery-like foliage, fennel has an aniseed flavour. The fleshy bulbous stem, seeds and foliage are often used in fish dishes. For a smaller herb with a similar-flavoured foliage, try the annual dill.

ROSEMARY
Rosmarinus officinalis

An evergreen shrub that produces blue flowers. The silvery, aromatic foliage is often associated with lamb dishes. Rosemary comes in a variety of sizes and can be trailing or upright.

SAGE
Salvia officinalis

An evergreen shrubby herb with aromatic foliage, often used in savoury meat dishes. The usual form has light green foliage, but there are attractive purple and tricolour versions, too. It is an evergreen perennial and may need cutting back to half its size once a year to prevent it getting too large and woody.

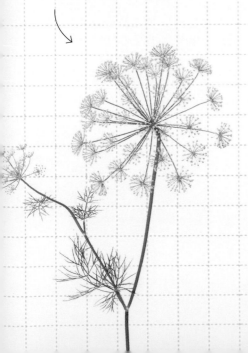

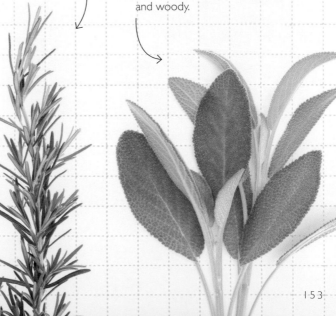

45

Grow a bonsai tree

Bonsai is the ancient Japanese technique of growing miniature or dwarf trees. Once you understand a bonsai's needs, it is easier than you might think to grow, and the resulting tiny trees make spectacular features in a garden or even inside.

Most trees and shrubs can have their natural size curtailed and be grown as a bonsai. It is important to start with young, small trees, as they are supple and pliable enough to manipulate into your chosen shape. Alternatively, you can buy a ready-trained bonsai tree from a garden centre, although this is a more expensive option.

One of the easiest and best trees for growing as a bonsai is Japanese maple (*Acer palmatum*); it produces lots of branches, so creating a nice shape is easy, and it grows moderately in the summer, so controlling the shape is manageable. Best of all, Japanese maples make beautiful specimens, with their trunk taking on an old, gnarly, ancient look. Their attractive five-lobbed foliage displays a spectacular range of colours in autumn.

Container your plant

Bonsai trees need to be grown in containers to curtail their growth. Plant them in a 50:50 mix of loam-based peat-free compost and sharp sand. Containers for bonsai are often shallow, sometimes even trays. Aim for a container that is one-third the height of the tree. It may be necessary to trim the roots first to make it fit.

Pruning, training and pinching

The art of restricting the tree's size is in the pruning and training. You can shape the tree how you wish by paying careful attention to where the branches are and removing the ones that are not suitable for your desired shape.

Removing some of the thicker branches, to give the tree its formative shape, should be carried out in winter. Avoid pruning Japanese maples in spring as they tend to 'bleed' sap.

As the tree grows during the spring and summer, some of the more vigorous-growing tips can be pinched back by about a third. This will encourage the growth of more shoots further down, giving them a denser canopy.

Wiring

Wrapping wire around branches can further control the tree's shape. There is no right or wrong to where the branches are positioned, it is just down to how you want the tree to be shaped. Have a play with using wires on the branches to try and manipulate where and how they are growing.

Clockwise from above left: Mature bonsai survive in the shallowest of trays, which will keep the plant compact.

Azaleas and other rhododendrons make spectacular bonsai displays when in flower.

Acers (Japanese maples) have beautiful foliage and an attractive habit, making them great subjects for bonsai growing.

To keep a bonsai compact it needs regular training and pruning.

Be inspired by a Japanese garden

Japanese-style gardens are very popular due to their simplicity and beauty. They are designed for contemplation and relaxation, with every element placed carefully to enhance the overall design. Perhaps not everybody has room for tea ceremony houses, elaborate bridges and cascading waterfalls, but many features can be incorporated on a smaller scale.

Many elements in a Japanese garden are symbolic, representing the surrounding landscape: rocks and boulders represent islands; ponds represent lakes and seas; plants represent the woodlands and nature. This is a key concept to bear in mind as you create your own version of a Japanese garden.

Choose your plants

There are plenty of Japanese plants to choose from, and here are just a few popular choices:

- **Japanese maples (*Acer palmatum*)** – are medium-sized trees with beautiful leaves that change colour in autumn and are typically featured in almost all Japanese gardens.

- **Upright bamboos** – are also popular, and add texture and movement as they sway in the wind. If planting them in the ground, choose a clump-forming type and one with slow growth. Black bamboo (*Phyllostachys nigra*) is an attractive black-stemmed bamboo; for a golden stem, try fish-pole bamboo (*Phyllostachys aurea*).

- **A flowering cherry tree** – can be planted, if room, for enjoyment of its spring blossom. The Japanese love cherry blossom so much that they have festivals each year to celebrate, called *Hanami*. There are many varieties to choose from, and one can be grown in a pot if space is limited.

- **Azaleas and rhododendrons** – are evergreen shrubs producing bright flowers in spring. They prefer acidic soil but can also be grown in pots of ericaceous compost if the soil is unsuitable.

Position your plants

Whereas many other types of garden pack as many plants as possible into an outdoor space, Japanese planting is different. The positioning of each individual plant is considered and deliberate. Many plants tend to be planted alone as a single feature, so their natural form can be admired from all sides, or in small clusters of identical or similar plants.

Japanese techniques

Cloud pruning – if you want the authentic Japanese look, then you could try 'cloud pruning'. This is an ornamental pruning technique involving shaping the branches of foliage into round balls that look like drifting clouds. Grow small-leaved evergreen shrubs such as Japanese holly (*Ilex crenata*) and Japanese privet (*Ligustrum japonicum*).

Hard landscaping – often, the surface of a Japanese garden is covered in gravel. This gives it a unifying design, linking all the individual elements together. As well as looking good, it is easy to maintain and helps to suppress weeds. Rocks of various sizes are also featured in a Japanese garden. Simplicity and beauty are key, so place rocks strategically to look natural and free from clutter. Lay a weed-suppressing membrane over the ground first, before adding the gravel to a depth of 5cm (2in).

Watery reflections – water is a key feature in most Japanese gardens, designed to move the senses. Reflections from the water instils a sense of symmetry and balance in the landscape. A still pond can evoke serenity and contemplation, while moving water adds a natural flow to the mood. Koi carp could be added to the water, their bright orange, white and red colours creating a contrast between the dark pool in which they swim. If there is space, a small bridge linking one side to the other adds symmetry. It also provides a position to look directly down into the water, for contemplation and to observe the clouds and sky reflections in the water.

Reseed or turf a lawn

Lawns make a beautiful, natural, verdant feature in a garden. Their neutral green colour acts as a soft background for brighter plants, while their soft texture provides a comfortable surface for sitting or playing on. Lawns can be created in the smallest of gardens, and can form creative shapes to enhance the overall design.

When establishing a lawn there are two options: you can sow the area with lawn seed or you can lay pre-established rolls of turf. Here are some of the pros to both approaches, so you can decide which might work best for you:

Benefits to sowing seed

Cost – a box or bag of grass seed is much cheaper than using rolls of turf. In fact, it is estimated that buying turf is ten times more expensive than the equivalent amount of seed. Delivery also costs less due to the lightness of seed packages.

Less back-breaking – if you struggle with fitness or mobility, you might find carrying a number of heavy rolls of turf into the back garden hard work. A packet of seed is much lighter to sow.

No rush – when rolls of turf are delivered or collected, there is only a small window of opportunity of a day or two before they need to be laid. After this, they will start to perish. This isn't ideal if the weather takes a turn for the worse, or if your plans change at short notice. A packet of seed, on the other hand, can last for a couple of years, so you can sow it when you're ready.

Choice – the type of grass mix available with rolls of turf is limited, whereas there is far more choice when it comes to selecting seed mixes. This is useful if you have a tricky area to cover, such as heavy clay, partial shade or acidic conditions, where common turf mixes might not work.

Benefits to laying turf

Time – laying a lawn is much quicker to establish than sowing seed. Turf transforms an area instantly, making it look lovely and green as soon as it is laid. Within a week or two it can be walked on. The lawn will also smother any emerging weeds. Seed, on the other hand, can take weeks to establish, and in the meantime, weeds may germinate in the soil and compete with the grass.

Ease – laying turf is reasonably easy and success rates are high, so long as they are kept watered regularly after laying. Also, the preparation of the soil is easier for laying turf than for seed sowing. The lawn only needs to be coarsely raked level for turf laying, whereas seed sowing requires a fine, crumbly soil for it to germinate.

All year round – turf can be laid almost all year round except for extreme periods of drought or freezing cold weather. Seed will only germinate between spring and late summer, when the temperatures are warm enough.

Slope cover – turf is easier to establish on slopes than seed, which can wash away in the rain or when watered. On steep slopes, pegs can be used to hold the turf in place.

Birds – if an area has been seeded then a vigilant watch for seed-loving birds will be needed. The area will probably need netting, but even then, birds have an amazing capacity to get under it. Turf does not have this problem, as the seed is already germinated.

From top: Avoid walking on a lawn that has been sown recently with grass seed until it is established.

Make sure each roll of turf is butted up closely to the next one, to ensure the edges do not dry out.

Once the turf is laid, water it well, unless rainfall is predicted soon afterward.

How to lay turf

Laying turf in a garden immediately transforms it into something lush and green. It can be laid at most times of the year and can be walked on just a week or two after laying. (For more vigorous activities, it is worth waiting a few more weeks until the roots from the turf have 'knitted' into the soil below it.)

You will need:

Fork

x1

Turf

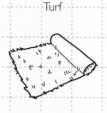

Organic matter such as compost or rotted manure

Half-moon edger

x1

Rake

x1

❶ Prepare the area by lightly digging it over with a fork. Add organic matter into the soil and rake it level, removing any perennial weeds or large stones. Leave the soil to settle for a few days, otherwise the level might sink once the turf has been laid.

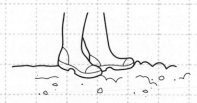

❷ Firm the soil down to knock out any large air pockets. This can be done by shuffling your feet along the surface slowly, making sure you distribute your weight evenly with each footstep. Then give the soil a final rake over, to ensure that the entire surface is smooth and level, and that any stones or lumpy bits are removed.

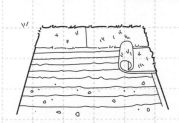

3 When it comes to laying the turf, it is best to start by laying the first row along the back edge, to avoid constantly walking across it. Continue to work along the next row, and gradually work your way to the front.

Keys to success

• If you end up being waylaid and can't lay your turf immediately, you should roll out your turf and water it each day until you are ready.

• Avoid leaving a short length of turf at the end of a row as it can dry out easily. Instead, place a longer, full length at the end and fill in the gap in the middle with the shorter length.

• In dry weather, the lawn will need watering each day until it is established.

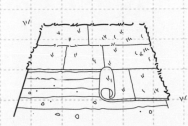

4 When starting each row, it is best to stagger the turf so that the joints are not lined up with the start of the row before it – a bit like brickwork. This is to create a stronger bond between each row.

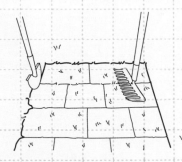

5 Lightly firm down each roll of turf by tamping down with the back of the rake. When the area has been covered, the ends can be trimmed off with a half-moon edger for a nice finish.

48

Create paths and paved areas

More than a practical solution for getting from one area to another in the garden, paths are an integral part of any garden design. They provide a structure to the overall space and link various areas of the garden together. Even in a small space, a path can be used to lead the eye toward focal points and areas of interest.

Paths can be made from a variety of different materials, depending on budget, resources and DIY skills. These include stone, slate, gravel, lawn, patio slabs, bricks, concrete, tarmac or woodchip. The material of the path can set the tone for the garden theme or design. Ideally, the material and style of the path should match or be in keeping with the surrounding area. For example, a formal herringbone brick would suit a formal setting; conversely, a woodchip path would be appropriate for a woodland garden.

Woodchip paths

Paths constructed out of natural materials usually look best in an informal situation. Woodchip paths are suitable for allotments, cottage gardens and other natural settings. One of the benefits of this type of path is that it requires practically no DIY skills. Plus, if tree surgeons are working nearby in the area, they are often happy to give away the wood chip from their shredder/chipper. If not, wood chippings or bark can be bought online and from most garden centres.

Lay down landscape fabric over the path surface to stop the woodchip mixing into the soil. Rake the wood chippings level to a depth of 6cm (2in). Complete the woodland theme by edging the path with sections of branches laid on their sides. Top up the path with chippings every couple of years.

Stepping stones

Another simple method of creating a path is to use stepping stones. These can be wooden log off-cuts, stones or patio slabs. One of the benefits is that in a small garden you hardly lose any space to paths, as the stepping stones can thread through planting areas or lawn. They are also easy to lay – simply dig out a gap in the soil and slot them in. If placing them in the lawn, they should be placed flush or below the surface so they can be mowed over easily. Be aware that wooden stepping stones can be slippery; cut 5mm (⅛in) grooves into the surface every 2cm (¾in) across to improve grip, or fix chicken wire to them with staples.

Gravel paths

Blending nicely into both formal and informal gardens, gravel is reasonably cheap compared to bricks or patio slabs, but if you are planning on using a wheelbarrow regularly on it, you may wish to reconsider, as it can be hard work pushing one on this surface. It is also unsuitable for wheelchair users.

Bricks and patio slabs

Slabs and bricks make the most hardwearing paths, but also the costliest. However, if a path is going to be used regularly, then it is worth investing in something that is durable, practical and is going to look good.

Clockwise from opposite: An informal, natural woodchip path is easy to install and looks great in a cottage garden.

Stepping stones are a simple solution for creating an impromptu path across flowerbeds and lawns.

Herringbone brickwork looks impressive and is long lasting.

Patio slabs can be painted black and used in contemporary gardens to dramatic effect.

Build a rustic brick path

A rustic brick path not only looks good but also provides a solid structure on which to walk. New bricks can be used, but old, upcycled bricks have far more character. They can often be found in reclamation yards, or even being thrown away in a skip. This path is simple to make, requiring only basic DIY skills. Best of all, there is no need for any messy mixing of cement to hold the bricks together.

You will need:

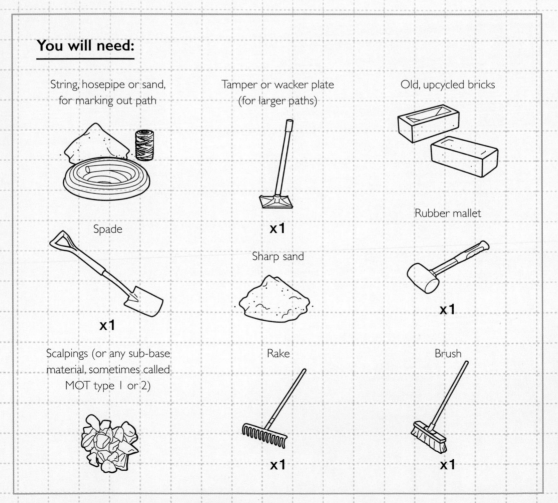

String, hosepipe or sand, for marking out path

Tamper or wacker plate (for larger paths)

x1

Old, upcycled bricks

Spade

x1

Sharp sand

Rubber mallet

x1

Scalpings (or any sub-base material, sometimes called MOT type 1 or 2)

Rake

x1

Brush

x1

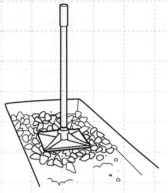

1 Mark out the path using string, a hosepipe or a trail of sand. Dig out the area using a spade to a depth of 15cm (6in) plus the depth of your bricks. Allow for 1.5cm (½in) for pushing the bricks into the sand.

2 Fill the trench with scalpings to a depth of 10cm (4in) and use a tamper to firm them down. For larger paths, it might be worth hiring a wacker plate to compact the scalpings.

3 Add a 5cm- (2in-) deep layer of sharp sand. Rake it level and use an off-cut of wood to tamp it down firmly, but leaving it soft enough so that bricks can be pushed into it.

4 Start laying the bricks by pushing them 1.5cm (½in) into the sand. Work across the path to the desired width. Stagger each new row to create a more cohesive structure. Use a rubber mallet to tap each one down, so they are level with one another.

5 Once the bricks have all been laid, brush more sharp sand between them to provide more stability in the path, and also to provide drainage. More sand may need to be swept in once a year to replace any that gets washed away.

49

Sow a mini wildflower meadow

Creating your own patch full of wildflowers can look spectacular. Not only will you enjoy the colourful flowers, but the wildlife will thank you for it, too. One of the many other benefits of a mini meadow is that it is easy to achieve, and the effects are almost instant. Within a few weeks of sowing seed, there will be a colourful display of flowers.

You don't need much space to create a mini flower meadow. A 1x1m (3x3ft) patch is fine, although if you do have more space, even better. If you have less space, sow into a container or raised bed.

Choose your seed mix

The first decision to make is whether you wish to sow perennial or annual seeds. Generally, annual seeds prefer fertile soil and should be sown in spring, whereas perennial seeds can be sown in autumn and prefer infertile soil, because otherwise grass tends to compete and swamp the wildflowers.

There are lots of both annual and perennial mixes available from seed companies. Do check that they are suitable for the conditions of your garden or growing space.

A blooming cornfield

An annual cornfield mix is popular for an authentic-looking mini wildflower meadow. Plants in the mix usually include cornflower, corn poppy, corn marigold and corn cockle.

How to sow

Dig over the soil, remove any weeds, then rake it level. Choose a calm day to sow the seeds, as they are light and can easily blow out of your hand and land elsewhere on windy days. Check the sowing rates on the packet, then scatter the seeds accordingly. Rates can often be as low as 1g per square metre, so to make it easier, it is worth mixing each square metre amount

Yellow rattle

If sowing a perennial mix, it is worth including some yellow rattle (*Rhinanthus minor*). These are parasitic wildflowers that help to deplete competing vigorous grasses of their energy, giving the wildflowers a fighting chance.

with silver sand for a better distribution. If sowing a large area, mark out 1x1m (3x3ft) square grids using canes and string, so that the correct rates can be applied to each square.

For an even distribution, sow half the amount in the square in one direction, and the other half at 90 degrees to that. After sowing, rake the seeds in lightly so they are just below the surface of the soil.

Aftercare

In small areas, place a net over the seeds to protect them from birds; for larger areas, hang old CDs or strips of silver foil on a line between two bamboo canes above the sown areas. The bright flashing reflections from the sunlight catching the metallic material helps to deter the birds.

Once annual flowers have finished flowering they should be left for a few weeks to allow the wildlife to enjoy the seeds. Delaying cutting them down also gives some of the seeds an opportunity to drop to the ground, meaning they might grow and flower the following year.

Complementary flowers

To complement autumn shrubs, try planting low-growing bulbs at the base of them. This can be done for plants in containers or directly in the ground. Meadow saffron (*Colchicum autumnale*), saffron crocus (*Crocus sativus*) and autumn-flowering cyclamen (*Cyclamen hederifolium*) make brightly coloured displays.

50

Add splashes of autumn colour

A display of brightly coloured autumn leaves brightens up the darkest of days as winter approaches. Leaves start to change colour as temperatures drop, and this is when all those fiery hues start to appear, marking the change of seasons in the garden.

Many of us associate autumnal colour with fiery displays of colourful foliage as the growing season draws to a close. There are lots of trees and shrubs with showy autumn foliage to choose from, and all of them can be grown in a container if there isn't room to plant one directly into the soil.

Place containers next to one another, with taller ones toward the back. The different shades of colours will create dramatic contrast, with all sorts of hues on display, including bright red, orange, yellow, gold and purple.

The brightest displays

If you only have room for one tree with beautiful autumn foliage, then Chinese dogwood (*Cornus kousa* var. *chinensis*) is a good choice, as in addition to colourful leaves in autumn, it has attractive white bracts in spring and large, bright red berries in late summer.

Clockwise from top left: Cotinus, or smoke bush, has dramatic and striking purple and red foliage as autumn approaches.

Colourful autumn berries extend the colour in a garden when some of the summer flowers have started to fade.

Sweet gum is a medium-sized tree with foliage that gives one of the best displays of autumn colour.

Smoke tree 'Royal Purple' (*Cotinus coggygria* 'Royal Purple') has purple foliage during the summer, with impressive plumes of flowers that look like balls of smoke, and in autumn the leaves turn the most extravagant bright red colour. If planted in the ground, it will get to a height of 5m (16ft), but its size will be curtailed if grown in a container.

If you are after a plant with autumn colour that can also provide you with fruit, then the blueberry is a perfect choice. It requires an acidic soil, so if this doesn't exist in the garden, then plant in a container in peat-free ericaceous compost. There are lots of varieties to choose from, so you can enjoy delicious juicy berries in summer and fiery crimson-red leaves in autumn.

For a slightly larger garden, sweet gum (*Liquidambar styraciflua*) is a medium-sized tree with impressively bright autumn foliage. An added bonus is that if you tear a leaf and smell it, it has a sweet, eucalyptus-like fragrance.

Katsura tree (*Cercidiphyllum japonicum*) is a similar-sized tree with stunning autumn colour. As the leaves change colour they give off a lovely aroma reminiscent of burnt candy or candyfloss, which fills the surrounding air.

Glossary

aerate Process used in lawn maintenance to add air to the roots and reduce compaction; usually involves pushing a fork about 5cm (2in) into the lawn at about 10cm (4in) intervals.

annuals Plants that grow, flower and die in one year.

friable Crumbly texture of soil that is ideal for sowing or planting in.

hardpans Compacted pieces of soil below the surface that can prevent roots from extending downward.

hardy Plants that can tolerate cold weather and frosts; suitable for growing outside all year round.

herbaceous perennials Plants made of soft material, which die back below ground in autumn and reappear in spring.

mulch Material used to cover the ground to suppress weeds and retain moisture; can include compost, manure, woodchip, pebbles, slate or weed-suppressing membrane.

obelisk Tall, tapering garden structure, usually four sided, used to train climbing plants up.

perennials Plants that live for more than one year.

potager Name originating from France, literally means a 'soup mix', used to describe a type of ornamental kitchen garden.

propagator Box or tray that is used to help seeds germinate and for young seedlings to grow in. Some plants have electric, heated trays to provide bottom heat, while others are just insulated boxes.

scarify Technique used in lawn care to remove 'thatch', or dead grass, and moss from among grass blades, improving air circulation and reducing fungal diseases.

tender Plants that are susceptible to frosts and cold weather; should not be grown outside in winter.

tilth Crumbly soil texture, similar to *friable*, making it easy to sow or plant in.

Further resources

Books

100 Perfect Plants by Simon Akeroyd (National Trust, 2017)

50 Ways to Outsmart a Squirrel & Other Garden Pests by Simon Akeroyd (Mitchell Beazley, 2021)

The Good Gardener by Simon Akeroyd (National Trust, 2015)

RHS Perfect Compost by Simon Akeroyd (National Trust, 2020)

RHS Perfect Lawns by Simon Akeroyd (National Trust 2019)

RHS Shrubs and Small Trees by Simon Akeroyd (DK, 2008)

RHS Practical House Plant Book by Zia Allaway and Fran Bailey (DK, 2018)

RHS How Can I Help Hedgehogs by Helen Bostock and Sophie Collins (Mitchell Beazley, 2019)

RHS Plants for Places (DK, 2011)

RHS The Creative Gardener by Matt Frost (DK, 2022)

RHS How to Plant a Garden by Matt James (Mitchell Beazley, 2016)

RHS Grow Your Own Veg & Fruit Bible by Carol Klein (Mitchell Beazley, 2020)

RHS Complete Gardener's Manual by RHS (DK, 2020)

RHS Encyclopedia of Gardening Techniques by RHS (Mitchell Beazley, 2008)

RHS Companion to Scented Plants by Stephen Lacey (Frances Lincoln, 2016)

RHS Water Gardening by Peter Robinson (DK, 1997)

RHS Small Garden Handbook by Andrew Wilson (Mitchell Beazley, 2013)

Websites

www.rhs.org.uk

www.simonakeroyd.co.uk

www.gardenersworld.com

Index

A

Acer davidii 18
achillea 26
aconite 28, 92
African lily 36
aloe 131
alpine plants 61, 62–5
Amazon elephant ear 108
angel's trumpet 47
annual flowers 61, 116–19
annual meadow grass 26, 43
annual weeds 43
annual wildflower meadows 166
apple 94, 96–7, 120
apricot 94
aquilegia 93
arbours 55
archways 52, 84
armand clematis 86
artemisia 34
asparagus 61
assessing your space 8–11
astrantia 28, 93
autumn 168–9
autumn-flowering cyclamen 168
avocado 120
azalea 156

B

bamboo 55, 58, 156
banana plant 39, 55
bar room plant 126
bareroot trees 88
bark 49
basil 150
bathroom plants 124–7, 129
bats 22, 26
bedding plants 61
bees 41
beetles 26
beetroot 75, 79, 80, 102
bellflower 'Birch hybrid' 64
Benjamin tree 109
big blue lilyturf 93
bindweed 43
biodiversity 41
birch 48, 49
bird of paradise 107, 109
birds 22, 23, 26, 52, 159
black bamboo 156
black-eyed Susan vine 52
'black gold' 142
black mondo grass 93
blackcurrent 97
blackthorn 88
bleeding hearts 32
blueberry 97, 169
bonsai trees 154–5
bramble 43
brick paths 163, 164–5
broad bean 75, 78
bromeliads 107
buddleja 22
bug hotels 24–5
bulbs 18–21, 168
bush maidenhair fern 130
buttercup 28
butterflies 22, 23, 26

C

cacti 132–5
callicarpa 48
canna 39
carex 34, 93
carnivorous plants 124, 128–9
carrot 71, 75, 78, 80
caterpillars 26
celery 102
chard 102
cherry 94, 120, 156
chia seeds 104–5
chicory 102
Chinese dogwood 169
chives 28, 150, 152
chocolate vine 87
chrysanthemum 28
cider gum 18
cinnabar moth 41
clematis 52
climbing hydrangea 33
climbing plants 33, 52, 61, 82, 84–7
closed terrariums 130–1
cloud-pruning 157
clover 41
cobweb houseleek 65
coleus 112
comfrey 41, 140–1
composting 138–9, 142–3, 144
containers
  autumn 168, 169
  bonsai 154
  fruit trees 94–7
  hanging 77, 98–9, 113, 114–15
  herbs 150–3
  shade-loving 92–3
  spring 18
  types 92, 94, 98, 100
  vegetables 73, 74–5
  vertical 76
  watering 61
  wildlife 22
winter 50–1
cordyline 39
coriander 102, 150
corn cockle 166
corn marigold 166
corn poppy 166
cornflower 166
cosmos 117
couch grass 43
courtyard gardens 12, 13, 68
cranberry 49, 98, 99
cranesbill geranium 93
cress 102
crocus 21
cucamelon 76
cucumber 76
curly parsley 153
cut-and-come-again crops 102–5
cuttings 112–15
cyclamen 18, 49, 92, 168

D

daffodil 20, 28
dahlia 39
dame's violet 47
damson 94, 120
dandelion 41, 43
daphne 28, 48, 49
delphinium 28
dog rose 54
dogwood 48, 49, 169
drought-resistant plants 34, 36–7
dry gardens 34–7
dwarf sweet box 49

Credits

I would like to thank Sorrel Wood, Sara Harper and Katie Crous from the Quarto Publishing team for being so wonderful to work with. Also, I owe a big debt of gratitude to Simon Maughan and the RHS team for their knowledge, advice and support. I would also like to thank Sarah Skeate for her beautiful illustrations, and Wayne Blades for design and picture research.

Picture credits

Adobe Stock 6–7 FollowTheFlow; 20L Yury Kisialiou; 20T ulkan; 21TR Soyka; 21BL antonel; 21BR firewings; 28(2) tonigenes; 28(3) robynmac; 28(5) yongkiet; 31 mashiki; 33TL uzuri; 33TR azure; 34–35 MXW Photography; 36L Richard Griffin; 37L Roman Ivaschenko; 39L marjancermelj; 40–41 Lili-OK; 57 Alene Pierro; 68 AlexanderDenisenko; 71 freebreath; 87R Fotolyse; 131 NesolenayaAleksandra; 133L Rawpixel.com; 135 warasit

Alamy 16 German Pineda; 38–39 Holmes Garden Photos; 46L Nature cutout's; 47BL Panther Media GmbH; 65M Christina Bollen; 95T Dorling Kindersley Ltd; 97 Alister Firth; 144 Pollen Photos

Dreamstime 54 Liewluck; 56 Laimdota Grivane; 63B Angelacottingham; 101 Jason Finn; 132 Saletomic; 157T Elena Milovzorova

Getty 77T mtreasure; 146 SolStock; 155TL rrodrickbeiler

iStock 29 CBCK-Christine; 33BR FactoryTh; 35 fotolinchen; 155BL lathuric; 155TR stocknshares; 155MR Imagesbybarbara; 157B MovieAboutYou; 163T jenjen42

Shutterstock 4B tete_escape; 9 onzon; 10–11 Sergey V Kalyakin; 15 vichie81; 17L L. Feddes; 17R Delovely Pics; 19TL AntonSAN; 19ML photka; 19TR Annie Shropshire; 19B Lois GoBe; 20B ER_09; 21TL Melica; 22 Ken Griffiths; 23T Erni; 23M Andi111; 23B Yvonne Griffiths-Key; 25 Peter Turner Photography; 27TL SomeSense; 27TR Mark R Coons; 27B jax10289; 28(1) Swapan Photography; 28(4) domnitsky; 28(6) Scisetti Alfio; 32L Tom Pavlasek; 32R sbgoodwin; 33BL ABIES; 34 Olga_Ionina; 36TR emberiza; 36BR RAJU SONI; 37TR InfoFlowersPlants; 37BR Shan 16899; 39R Nixholas Mike; 40 Peter Turner Photography; 41 Andrei Begun; 42 Jon Rehg; 43L Vikafoto33; 43R Katrin85; 45 Whatafoto;

46R unpict; 47TL Svetlana Foote; 47TR Julius Elias; 47BR Nadezhda Nesterova; 48TL Tamar Ramishvili; 48TR Peter Turner Photography; 48B Tatyana Mut; 51 InfoFlowersPlants; 51T Molly Shannon; 51BL Happy Dragon; 51BR Elena Elisseeva; 55L Viacheslav Lopatin; 55R Menno van der Haven; 57 Roman Nerud; 59TL Ariene Studio; 59TR nnattalli; 59B Joanne Dale; 60TL Wirestock Creators; 60TR Helen Pitt; 60B Niall F; 63T K.-U. Haessler; 64L Martina Osmy; 64TR StockPictureGarden; 64BR muroPhotographer; 65T de2marco; 65B panattar; 66 Angela Royle; 73T lavizzara; 73BL Stanley Dullea; 73BR Graham Corney; 74T Praiwun Thungsarn; 74B Imfoto; 75T Peter is Shaw 1991; 75B tottoto; 77B Carl Stewart; 78L eelnosiva; 78T Philip Kinsey; 78R Katrinshine; 79T Valentyn Volkov; 79M xpixel; 79B PIXbank CZ; 80L OhSurat; 80R Catherine Eckert; 81 M GI; 83 Dmytro Balkhovitin; 85T Peter Turner Photography; 85L jonesyinc; 85R Trialist; 86L DariKor; 86T Evan Hutomo; 86R Tamara Kulikova; 87T Picture Partners; 87L Siriporn-88; 89T Jamie Hooper; 89L Manfred Ruckszio; 89M Dmitriev Mikhail; 89R Savanevich Viktar; 90 high fliers; 91 Margo K; 93TL claire norman; 93 Jenell Kasper; 93 Michael Warwick; 93 Zuzha; 95M seaonweb; 95R Steven Nilsson; 98L Robyn Gwilt; 98R stevemart; 99L Leslie Shields; 99R Natalia Korshunova; 102 Makina Alesya; 103 Evgrafova Svetlana; 105 The natures; 106–107 Followtheflow; 108T mokjc; 108B Reda.G; 109TL Mr.Teerapong Kunkaeo; 109TR xavirm21; 109BL deckorator; 109BR Scisetti Alfio; 110 liloon; 111 evgeniykleymenov; 112L ellinnur bakarudin; 112R Zainul Yudharta; 113T AngieYeoh; 113BL Yangxiong; 113BM Oat_Pittiwat; 113BR Rungnapa4289; 115 LorenaEscamilla; 116–117 Anjo Kan; 117 Ian Grainger; 119 Yuds; 121TL Aliye Aral; 121TR Thijmen Piek; 121BL Nastya_Eso; 121BR Anakumka; 123 M. Unal Ozmen; 124 aprilante; 126L MT.PHOTOSTOCK; 126TR SUPAPORNKH; 126BR Jiri Sebesta; 127L nbldesign; 127TR Jotika Pun; 127BR tetiana_u; 128L Studio Barcelona; 128R Dian Munteanu Permatasari; 129L Angel Santana Garcia; 129R Chun's; 130–131 qnula; 133R Sheila Fitzgerald; 136 anat chant; 138 Ingrid Balabanova; 139 Tommy Lee Walker; 138–139 Alison Hancock; 141 Kelly Whalley; 142 Katherine Heubeck; 143 Ashley-Belle Burns; 147T NayaDadara; 147B Alison Hancock; 149 NayaDadara; 151T Anne Kramer; 151B pullia; 152L Madlen; 152M Nataly Studio; 152R HamsterMan; 153L romiri; 153M Scisetti Alfio; 153TR Maks Narodenko; 153ML Mr. SUTTIPON YAKHAM; 153BL Nattika; 157M Pierpaolo Pulinas; 159T John-Kelly; 159M/B Olga Aniven; 161 New Africa; 162 Hannamariah; 163M Kittima05; 163B AePatt Journey; 166–167 July photographer; 167 Tom Meaker; 168TL Alexander Denisenko; 168TR peony graden; 168BR Keikona

Wayne Blades 4T, 95L, 116, 165